Francisco de Sales Oliveira Filho
Francisco Hevilásio Freire Pereira

Organomineral fertilization in watermelon cultivation in the semi-arid region of Paraiba

AF667389

Francisco de Sales Oliveira Filho
Francisco Hevilásio Freire Pereira

Organomineral fertilization in watermelon cultivation in the semi-arid region of Paraiba

A fertilization strategy to benefit the family agro-ecosystem

ScienciaScripts

Imprint

Any brand names and product names mentioned in this book are subject to trademark, brand or patent protection and are trademarks or registered trademarks of their respective holders. The use of brand names, product names, common names, trade names, product descriptions etc. even without a particular marking in this work is in no way to be construed to mean that such names may be regarded as unrestricted in respect of trademark and brand protection legislation and could thus be used by anyone.

Cover image: www.ingimage.com

This book is a translation from the original published under ISBN 978-613-9-68904-0.

Publisher:
Sciencia Scripts
is a trademark of
Dodo Books Indian Ocean Ltd. and OmniScriptum S.R.L publishing group

120 High Road, East Finchley, London, N2 9ED, United Kingdom
Str. Armeneasca 28/1, office 1, Chisinau MD-2012, Republic of Moldova, Europe
Printed at: see last page
ISBN: 978-620-8-24089-9

Copyright © Francisco de Sales Oliveira Filho, Francisco Hevilásio Freire Pereira
Copyright © 2024 Dodo Books Indian Ocean Ltd. and OmniScriptum S.R.L publishing group

SUMMARY

ACKNOWLEDGMENTS

To God, for giving me the gift of life.

To my family, especially my parents, for all their dedication and for shaping me as a citizen.

To all those who are part of the Postgraduate Program in Tropical Horticulture at the Federal University of Campina Grande Campus Pombal, for the opportunity given to me and for having believed in my ability and commitment.

I would like to thank Professor Francisco Hevilàsio Freire Pereira for his presence at all times and for his valuable teaching and guidance.

To all the new friends I've made, for their presence and support at all times during the course and conclusion of our work.

To Capes for the scholarship.

GENERAL SUMMARY

OLIVEIRA FILHO, Francisco de Sales & PEREIRA, Francisco Hevilàsio Freire.

Organomineral fertilization in watermelon cultivation in the semi-arid region of Paraiba, 2018. 71p.

Producing quantity and quality without harming the environment or increasing production costs is the main objective of small and medium-sized farms. To this end, new technologies are needed in order to reduce dependence on the market for industrialized inputs, prioritizing the use of renewable resources acquired on the property itself. The aim was to evaluate the effect of the combined and isolated use of mineral and organic fertilizers on watermelon crops. The experiment was carried out in an area located at the Federal University of Campina Grande (UFCG), Campus Pombal - PB (6° 48" 16""S and 37° 49" 15""W), from June to September 2013. The design was in randomized blocks with four replications arranged in a 3x5 factorial scheme where, in factor (A), different concentrations of nutrients N, P and K were allocated (50, 100 and 150% of the NPK recommendation for watermelon) and in factor (B) five proportions of mineral and organic fertilizer (100/0, 75/25, 50/50, 25/75 and 0/100). The dose of NPK corresponding to 100% was 120 kg ha^{-1} for N, P and K respectively. The following were evaluated: gas exchange, dry mass of leaves, stem, fruit, total dry mass, leaf area, nutrient accumulation in leaves, stem, fruit and total, total production, titratable acidity, total soluble solids and the ratio between soluble solids and titratable acidity. The concentration of 150% of the NPK recommendation for the watermelon crop was the most efficient in increasing the physiological characteristics and the accumulation of dry mass in the stem, leaf, fruit and total; the concentration of 100% of the NPK recommendation does not adequately meet the nutritional needs of the "Olimpia" watermelon crop in the Paraiba Sertao; the combined application of mineral and organic fertilizer provides a photosynthetic rate equivalent to the application of mineral fertilizer alone, with the 50/50 ratio being the most efficient for this variable. The use of manure alone does not contribute to the increase in dry mass in the vegetative part, when compared to its combined use with mineral fertilizers, with the 75/25 and 50/50 proportions being the most efficient in the accumulation of dry mass in the watermelon tree; the concentration of 150% of the recommended dose was the most effective in the accumulation of NPK in the leaves, stem and fruit of the watermelon tree; the proportions of mineral and organic fertilizer 50/50 and 75/25 were more effective in the accumulation of N and P and 0/100 in that of K in the leaves of the watermelon tree; the accumulation of NPK in the fruit and vegetative part showed the following order of magnitude: K>N>P; the fruits were the preferential drains in NPK in relation to the vegetative part; the 150% concentration of nutrients was more effective in increasing productivity in the mineral and organic fertilizer proportions of 0/100, 75/25 and 25/75; the 100 and 150% concentrations were more effective in the accumulation of total soluble solids when combined with mineral and organic fertilizer proportions of 75/25 and 50/50.

Keywords: *Citrullus lanatus*, fertilizers, physiology, production, mineral nutrition

1. GENERAL INTRODUCTION

The watermelon tree, *Citrullus lanatus* (Thunb.) Matsum & Nakai, belonging to the Cucurbitaceae family, is a fruit-bearing vegetable that originated in Africa and is grown all over the world (CARVALHO, 2007). In Brazilian agribusiness, the cultivation of this cucumber is of significant economic importance, with a 4.7% share of the total value of fruit production, which places it sixth among the twenty-two most produced fruits in the country (IBGE, 2012).

Brazil occupies the fourth position in the ranking of the twenty largest watermelon producers in the world, but has the lowest yield of the group of countries that lead production (FAOSTAT, 2016). Watermelon production in Brazil in 2016 was 2,090,432 tons in 90,447 hectares of harvested area, which corresponds to a yield of approximately 23.11 tons ha^{-1} . The Northeast region is responsible for 30.89% of the country's fruit production, with Bahia, Rio Grande do Norte and Ceará being the top producing states, respectively. Paraiba is among the northeastern states with the smallest harvested area, 206 ha, and the lowest production, 4,002 t, ahead of the state of Alagoas, which has a harvested area of 82 ha and an average production of 2,200 t (IBGE, 2012).

Watermelon cultivation is often limited by the high cost of inputs recommended in the literature (FILGUEIRA, 2008). From this point of view, studies into new technologies aimed at reducing costs while maintaining or even improving productivity are important, especially in the semi-arid regions of Paraiba, where few producers grow this crop, despite the favorable soil and climate conditions.

The success of watermelon cultivation is strongly influenced by the fertility of the soil in the area where it is grown and, in order to raise soil fertility levels, producers have resorted to applying mineral fertilizers. These, however, represent a significant portion of production costs. By August 2010, the fertilizer industry had sold around 2.7 million tons of fertilizers (ANDA, 2010). The cost of mineral fertilizers and growing environmental pollution have led to an increase in demand for research to assess the technical and economic viability of using organic waste (MELO et al., 2008).

Brazil ranks 4th among the countries that consume mineral fertilizers, accounting for 6.25% of the global total, with China, India and the United States being the biggest consumers. However, the country only stands out as a major producer of phosphate, ranking 6th in the world, with production of around 6.3 million tons of concentrate in 2010, which represents 3.6% of the estimated world production of 160 million tons (IBRAM, 2012). Data on Brazil's production and imports of N, P and K fertilizers show its high dependence on foreign suppliers, especially nitrogen and potassium fertilizers, where 75% of nitrogen and 92% of potassium are imported (ANDA, 2010).

In recent years, there has been a growing interest in the production of organic fertilizers, mainly due to the search for soil management alternatives with an organic approach and with characteristics that differ from the intensive use of industrialized chemical fertilizers (SIMÔES et al., 2007). The use of organic materials in agricultural systems has become widespread, given their important economic and environmental contributions. From an economic point of view, the main factor is the increase in the producer's income, due to the reduction in the use of industrialized fertilizers and pesticides, which makes cultivation viable for less capitalized farmers. Concern about the harmful effects on the environment caused by the dominant model of agriculture has led to the emergence of new technologies for reusing waste, whether urban, industrial or agricultural, in order to clean up the environment and serve as alternatives to the use of industrialized fertilizers in agriculture.

Within this agricultural scenario of dependence on the market for agricultural inputs, especially industrialized mineral fertilizers, the adverse effects they cause to the environment and the socio-economic profile of farmers in the Northeast region, we propose to carry out this work whose objective was to evaluate the effect of using organic fertilizer combined with industrialized mineral fertilizers on the agronomic and fruit quality characteristics of the watermelon tree.

2. THEORETICAL FRAMEWORK

2.1 Mineral fertilization

Mineral fertilizers are essential and indispensable inputs for the conventional agricultural model, meeting both agricultural development and quality goals, but they must be used responsibly. Malavolta et al. (2002) define mineral fertilizers as products of an inorganic nature, natural or synthetic, which supply nutrients to plants. They can be found as simple mineral fertilizers, which are divided according to the species of the main nutrient they contain, or in mixtures, so that they can be applied together in the same operation. According to Faquin et al. (2007), mineral fertilizers are made up of inorganic chemical compounds and can be made up of synthetic or artificial organic compounds.

The use of inorganic fertilizers in the agreste and semi-arid region of Paraiba is infrequent due to the limited purchasing power of low-income producers, difficult access to agricultural credit and high variability in rainfall (SILVA GALVÂO et al., 2008). Increases in production, provided by the use of fertilizers, must be accompanied by an increase or maintenance of the quality of the fruit produced (GRANGEIRO; CECiCILIO FILHO, 2004; MEDEIROS, 2008). It is known that mineral nutrition has a significant influence on the quality of watermelon fruit, but the search for solutions to the challenge of increasing production almost always affects the quality of the fruit.

The sources of the macronutrients N, P and K are the most important from the point of view of the production process. The other macro and micronutrients, despite their biological importance, have no economic expression in the fertilizer industry, nor significant commercial value, as they are used in very small quantities (FERNANDES; DIAS, 2006). The two nutrients most used as fertilizers in agriculture are nitrogen and phosphorus, the former due to its high demand by the plant and the latter due to its high fixation in soils (FAQUIN et al., 2007). Among the sources of nitrogen fertilizers, industrial urea is the main source of nitrogen used in agricultural production units due to the lower costs of obtaining it and the high concentration of N (46%), as well as the fact that it is chemically no different from animal urea (URQUIAGA; MALAVOLTA,

2002). N is subject to a large number of processes, especially transformations from organic to inorganic forms and vice versa, which can result in losses or gains for the system as a whole (RAIJ, 1991). It is estimated that only 1/3 of the nitrogen and potassium fertilizers incorporated into the soil are used by the plants, with the other part being lost through leaching, surface runoff and volatilization (ALFAIA, 1997).

The rising costs of nitrogen fertilizers, coupled with high losses in the field, demand management practices that result in high efficiency in nitrogen absorption by crops. Environmental risks are also taken into account when managing N in agricultural systems, since this nutrient is subject to high losses through erosion, leaching, denitrification and volatilization. Therefore, the ideal management of nitrogen fertilization should be defined as one that allows the crop's needs to be met, but with the minimum risk to the environment (FERNANDEZ, 2006).

Among the potassium fertilizer sources, KCl is the most widely used, due to its lower price and greater availability on the market, but some precautions should be taken before using it, such as: the high salt content which can damage germination, the root system and, consequently, plant development, the increased risk of soil salinization and the possibility of chlorine phytotoxicity (GRANGEIRO; CECILIO FILHO, 2004). Studies carried out by Silva et al. (2001) on the effects of potassium chloride on soil cultivated with chili peppers showed an increase in the electrical conductivity of the soil solution as the dose of this nutrient increased, which could compromise the development of the root system and the development and production of the crop.

According to Raij (1991), of the N, P and K fertilizer sources, P is the most used in fertilization in Brazil, because, according to Santos (2010), when phosphate fertilization is carried out, it is expected that the P will be available to the plants. However, the smallest part of the added P, around 10%, is in equilibrium with the P in solution, the other 90% forming non-labile P, which is not useful for the immediate growth of the plant. Although it is the nutrient most used in fertilization in Brazil, phosphorus is required less than nitrogen and potassium by watermelons. According to Epstein and Bloom (2006), this is due to the low availability of P in tropical soils, which occurs in the majority of Brazilian soils due to the high immobilization power of the added nutrient.

Work has been carried out in some Brazilian states to study the effect of applying nitrogen, phosphorus and potassium on watermelon production. Andrade Junior et al. (2006) when studying the effect of doses of 0 to 160 Kg ha^{-1} of N on the watermelon tree in Piaui observed an estimated yield of 60.17 t ha^{-1} at a dose of 97.61 Kg ha^{-1} . Gonçalves et al. (2011) when studying the effect of different doses of N and K on watermelon in Sao Paulo observed that the doses 79.8 Kg ha^{-1} of N and 88.5 Kg ha^{-1} of K were the most efficient in watermelon production, providing a commercial yield of 31,000 Kg ha^{-1} . Cecilio Filho and Grangeiro (2004) when evaluating watermelon productivity as a function of K sources and doses observed that the highest yield per plant was obtained with 132, 193 and Kg ha^{-1} of K2O ha^{-1} using K2SO4, KNO3 and KCl as sources, respectively. Freitas Junior et al. (2008) in their studies with different doses of phosphorus (0 to 360 Kg ha^{-1}) in the Congo watermelon hybrid in Cassilândia (MS), observed that the dose of 360 Kg ha^{-1} was responsible for the highest yield, 41.95 Kg ha .$^{-1}$

2.2 Organic fertilizer

The use of organic materials as a source of alternative raw materials for fertilizer production is a strategic measure from an environmental point of view, and is convenient as long as it is also viable from an economic point of view (FERNADES et al., 2003). Among the organic materials that can be used in agriculture, the use of animal manure occupies a prominent place with a positive effect on water infiltration and retention and an increase in the cation exchange capacity of soils (HOFFMANN et al., 2001).

Organic fertilization with cattle manure is an ancient practice, but with the introduction of industrialized fertilizer sources with high solubility and nutrient concentration in the mid-19th century, it lost its importance. Only in recent decades has organic fertilization regained its prestige, with growing concern for the environment, healthy eating and the need to give an appropriate destination to the large quantities produced in some countries (SALAZAR et al., 2005). Bovine manure has been widely used as a source of organic matter for the soil and nutrients for plants, making it an excellent alternative to mineral fertilizers (RODRIGUES et al., 2008). According to the same author, several researchers have carried out work using cattle manure as a substrate for the development of various species, mainly horticultural crops.

Manure has been used as an alternative for supplying nutrients, mainly nitrogen and phosphorus, in family farming areas in the semi-arid and harsh regions of northeastern Brazil (MENEZES; SALCEDO, 2007). According to Silva Galvao et al. (2008), cattle manure can both be a source of income for the farmer, in the case of selling the product, and a means of replenishing the soil with nutrients removed by the crops, since their studies showed that the amount of nutrients added annually by the manure exceeds the requirements of the crops and results in significant accumulations of C, N, P, K, Ca and Mg in the 0 to 20 cm layer. According to Marques (2006), animal manures are the most important organic fertilizers, due to their composition, relative availability and application benefits.

Increasing the use of organic fertilizers in relation to mineral fertilizers makes the agro-ecosystem more sustainable by improving the soil's physical, chemical and biological factors and consequently reducing the negative effects generated by intensive agricultural practices on the environment. Matos et al. (2008) concluded in their studies that organic fertilization, over the years, promotes an increase in total organic carbon, aggregate stability in water, P and N contents in the different aggregate classes, and decreases the C/N and C/P ratios.

According to Malavolta et al. (2002), the use of organic fertilizers not only favors soil drainage and aeration, but also increases water retention, nutrient levels and the population of beneficial organisms in the soil and on the plant, improving root development. The same author points out that it is impossible to replace mineral fertilizers with organic fertilizers due to the low concentration of nutrients in these materials.

Organic fertilization is an important source of nutrients, especially N, P, S and micronutrients. It is the only form of N storage that does not volatilize and is also responsible for 80% of the total phosphorus found in the soil (PIRES; JUNQUEIRA, 2001). The organic production model makes it possible to achieve good levels of productivity while avoiding the risks of chemical contamination for farmers, consumers and the environment (BORGES et al., 2013).

Organic fertilizers can be used as an alternative to mineral fertilizers to improve soil structure and increase microbial biomass (DAUDA et al., 2008). The need to use renewable forms of energy has revived the use of organic fertilizers around the world.

Cavalcante et al. (2010) evaluated different doses of cattle and goat manure in watermelon cultivation and observed that, regardless of the source, the dose of 10L per plant provided a similar number of fruits to commercial systems. Cattle manure is apparently the material most commonly used as an organic fertilizer (CANELLAS et al., 2005).

Silva et al. (2004), when evaluating the influence of this material on the yields of green ears and grains of maize, observed that manure influences the available water content and the maintenance of soil moisture, with linear increases in these characteristics, as well as a linear increase in the P content in the soil, thus increasing the number of marketable ears and the grain yield.

2.3 Organomineral fertilization

Organomineral fertilizer is a new and alternative product, the result of enriching organic fertilizers with mineral fertilizers. As a result of the higher concentration of nutrients compared to organic fertilizers, it has the advantage of being able to be used in smaller quantities per area, as well as lower transportation costs (FERNANDES; TESTZLAF, 2002).

According to Liu et al. (2009), the combined use of organic materials with mineral fertilizers is fundamental to developing more sustainable fertilization strategies. Decree 86.955 of 18/02/1982 included the term organo-mineral fertilizers in a law for the first time, defining it in Chapter I of the preliminary provisions as a fertilizer "resulting from a mixture or combination of mineral and organic fertilizers" (BRASIL, 1983).

Work has been carried out in some Brazilian states to study the effect of the combined use of mineral fertilizers on various crops. Fernandes et al. (2003) concluded that the use of organo-mineral fertilizers in the fertigation of the melon crop provided greater productivity and fruit of excellent quality when compared to fertilization with mineral fertilizers alone. Silva et al.(1999) observed that replacing part of the mineral fertilizer (formula 4-14-8) with organic compost from the decomposition of sugar cane bagasse and pig manure led to an increase in the number of leaves per plant, length of the main branch and fruit yield of hybrid II pumpkin, with the maximum yield being 13.60 t ha^{-1} obtained with the combination of 6.4 t ha^{-1} of organic compost and 360 kg ha^{-1} of the 4-14-8 formula.

Leao et al. (2008), who studied different levels of organic and chemical fertilization, observed lower yields of the Crimson Sweet variety of watermelon when these were used alone. Leite et al. (2003) reported that the use of animal manure combined with mineral fertilization has been an important strategy for improving soil fertility. Mueller et al. (2013) in their studies with organic fertilizer, with and without mineral fertilizer, in tomato cultivation, observed higher commercial yields obtained with the application of mineral fertilizer alone or with the application of organic fertilizer complemented with mineral fertilizer.

The addition of chicken litter increases the levels of K and S in the soil (CARVALHO et al., 2011). According to Andreola et al. (2000) the nutrient observed to accumulate most in the soil as a result of the use of manure is potassium, followed by calcium and magnesium. In their studies, they concluded that the use of organic fertilizer caused an accumulation of nutrients in the soil, while organomineral and mineral fertilizers showed a tendency to reduce the levels of potassium in the soil.

The incorporation of organic matter (OM) into the soil, which is capable of rapidly adsorbing applied phosphorus (P) in the form of fertilizer, increases the availability of P (GUPPY et al., 2005). The same authors state that this effect has been commonly referred to as competition between the decomposition products of OM and local adsorption of P by the soil, resulting in soil solutions with high concentrations of P. For Moreira and Siqueira (2006), the addition of organic material favours the microbial solubilization of phosphate, and this effect is related to the nature of this material. These authors also state that the solubilization of P in the soil can result from the production of CO_2 and organic acids from the mineralization of organic carbon and the production of enzymes and complexing chelating compounds by the microbiota. Among other attributes, there is a reduction in the maximum P adsorption capacity (SOUZA et al., 2006).

Bertol et al. (2010) in their studies on phosphorus loss via surface runoff in no-till systems under organic and mineral fertilization observed that the concentration of this nutrient in the soil was favoured by the application of organic fertilization compared to mineral fertilization. Borges et al. (2013) observed that organic fertilization favors the accumulation of P in jambu plants compared to mineral fertilization, with the recommended dose being 10 kg m^{-2} of farmyard manure.

BIBLIOGRAPHICAL REFERENCES

ALFAIA, S. S. Fate of nitrogen fertilizers in a yellow soil cultivated with Caupi Beans. **Acta Amazonas**, Amazonas, v. 27, n. 2, p. 65-72, 1997.

ANDA (National Association for the Dissemination of Fertilizers). **Statistical yearbook of the fertilizer sector**. Sao Paulo, 2010.

ANDRADE JÛNIOR, A. S. et al. Watermelon fruit production and quality after nitrogen application via fertigation. **Revista Brasileira de Engenharia Agricola e Ambiental**, Campina Grande, v. 10, n. 4, p. 836-841, 2006.

ANDREOLA, F. et al. Chemical properties of a structured purple soil, influenced by winter cover crops and organic and mineral fertilization. **Revista Brasileira de Ciências do Solo**, Viçosa, v. 24, p. 609-620, 2000.

BERTOL, O. J. et al. Phosphorus loss by surface runoff in no-till system under mineral and organic fertilization. **Scientia Agricola**, Piracicaba, v. 67, n. 1, p. 7177, 2010.

BORGES, L. S. et al. Productivity and accumulation of nutrients in plants of jambu, under mineral and organic fertilization mineral. **Ciências Agrârias**, v. 34, n. 1, p. 83, 2013.

BRAZIL. Ministry of Agriculture. National Secretariat for Agricultural Defense. **Inspection and supervision of the production and trade of fertilizers, correctives, inoculants, stimulants or biofertilizers intended for agriculture**. Brasilia: Secretaria de Fiscalizaçao Agropecuària, 1983. 86p.

CANELLAS, L. P.; BUSATO, J. G.; CAUME, D. J. **The use and management of humified organic matter from the perspective of agroecology**. In: CANELLAS. L. P.

& SANTOS, G. A. Humosfera: tratado preliminar sobre a quimica das substancias hùmicas. Campos dos Goytacazes, p. 244-267, 2005.

CARVALHO, L. C. C.; BEZERRA, F. M. L.; CARVALHO, M. A. R. Evapotranspiration and crop coefficients of seedless watermelon. **Revista Ciência**

Agronômica, Fortaleza, v. 39, n. 01, p. 53-59, 2007.

CARVALHO, R. E. et al. Mineral fertilizer and organic residue on soybean agronomic characteristics and soil nutrients. **Revista Ciência Agronômica**, Fortaleza, v. 42, n. 4, p. 930-939, 2011.

CAVALCANTE, I. H. L. et al. Organic fertilizers for watermelon cultivation in Bom Jesus-PI. **Revista Brasileira de Ciências Agrârias**, Recife, v. 5, n. 4, p. 518-524, 2010.

CECILIO FILHO, A. B.; GRANGEIRO, L. C. Watermelon crop productivity as a function of potassium sources and doses. **Ciência e Agrotecnologia**, Lavras, v. 28, n. 3, p. 561-569, 2004.

EPSTEIN, E.; BLOOM, A. J. **Nutriçâo mineral de plantas: principios e perspectivas**. Londrina: Editora Planta, 2006. 402p.

FAOSTAT. **Agricultural Data**. 2016. Available at <http://www.fao.org/faostat/en/#rankings/countries_by_commodity> Accessed on: 05/07/2018.

FAQUIN, V. et al. **Fertilizers and the Environment**. Lavras, UFLA/FAEPE, 2007. 86p.

FERNANDES, A. L. T.; RODRIGUES, G. P.; TESTEZLAF, R. Mineral and organomineral fertirrigation in relation to qualy of greenhouse cultivated melon. **Scientia Agricola**, Piracicaba, v. 60, n. 1, p. 149-154, 2003.

FERNANDES, A. L. T.; TESTEZLAF, R. Fertirrigation in the melon crop in a protected environment, using organomineral and chemical fertilizers. **Revista Brasileira de Engenharia Agricola e Ambiental**, Campina Grande, v. 6, n.1, p 45-50, 2002.

FERNADES, E.; DIAS, V. P. **Fertilizers: A global and systemic vision**. BNDS Sectorial. Rio de Janeiro, n.4, p. 97-138, 2006.

FERNANDEZ, F. C. S. **Nitrogen dynamics in the corn crop (*Zea mays* L.) in**

succession with black oats (*Avena strigos*) under no-till farming. 2009. 198p. Thesis (PhD in Soils and Plant Nutrition) - Escola Superior de Agricultura "Luiz de Queiroz", Universidade de Sao Paulo, Piracicaba, 2006.

FILGUEIRA, F. A. R. **Novo manual de Olericultura: agrotecnologia moderna na produçâo e comercializâo de hortaliças**. Viçosa, UFV, 2008. p. 342-348.

FREITAS JÙNIOR, A. N.; BISCARO, A. G.; SILVA, T. R. B. Phosphate fertilization in irrigated watermelon, in the municipality of Cassilândia (MS). **Revista Cultivando o Saber**, Cascavel, v. 1, n. 1, p. 1-6, 2008.

GALVÂO, S. R. S; SALCEDO, I. H.; OLIVEIRA, F. F. Accumulation of nutrients in sandy soils fertilized with cattle manure. **Pesquisa Agropecuâria Brasileira**, Brasilia, v.43, n. 1, p. 99-105, 2008.

GONÇALVES, M. V. I. et al. Leaf area index and yield of watermelon with seedless fruit as a function of plant spacing and N and K applied by fertigation. **Cientifica**, Jaboticabal, v. 39, n. 1, p. 25-33, 2011.

GRANGEIRO, L. C.; CECILIO FILHO, A. B. Accumulation and export of nutrients in seedless watermelon. **Cientifica**, Jaboticabal, v. 33, n. 1, p. 69-74, 2004.

GUPPY, C. N. et al. Competitive sorption reactions between phosphorus and organic matter in soil: a review. **Australian Journal of Soil Research**, Australia, v. 43, n. 2, p. 189-203, 2005.

HOFFMANN, I. A. et al. Farmers management strategies to maintain soil fert ility in a remote 11 ert in northwest Nigeria. **Agriculture, Ecosystems & Environment**, v. 86, n. 03, p. 263-275, 2001.

IBRAM. **Information and analysis of the Brazilian mineral economy**, 7ª Edition, 68 p, 2012.

BRAZILIAN INSTITUTE OF GEOGRAPHY AND STATISTICS. **Municipal agricultural production**. Temporary and permanent crops (2011). Rio de Janeiro, 2012, 97p.

LEÂO, D. S. S. et al. Watermelon productivity under different levels of chemical and organic fertilization. **Bioscience Journal**, Uberlândia, v. 24, n. 4, p. 32-41, 2008.

LEiTE, L. F. C. et al. Total organic carbon stocks and their compartments in clay loam under forest and under corn cultivated with mineral and organic fertilization. **Revista Brasileira de Ciências do Solo**, Viçosa, v. 27, p. 821-832, 2003.

LiU, M. et al. organic amendments with reduced chemical fertilizer promote soil microbial development and nutrient availability in a subtropical paddy field: the influence of quantity, type and application time of organic amendments. **Applied Soil Ecology**, v. 42, n. 02, p. 166-175, 2009.

MALAVoLTA, E.; GOMEs, F. P.; ALCARDE, J. C. **Adubos e Adubaçôes**, sao Paulo: Nobel, 2002. 200p.

MARQUES, L. F. **Production and quality of beet as a function of different dosages of bovine manure**. 2006. 37p. Monograph (Graduation in Agronomy) - Universidade Federal Rural do Semi-Àrido, Mossoró, 2006.

MATOS, E. S. et al. Stability of aggregates and distribution of carbon and nutrients in Argissolo under organic and mineral fertilization. **Pesquisa Agropecuâria Brasileira**, Brasilia, v.43, n.9, p.1221-1230, 2008.

MEDEIROS, D. C. **Production and quality of watermelon fertigated with N and K**. 2008, 70p. Thesis (Doctorate in Plant Science) Universidade Federal Rural do SemiÀrido, Mossoró, 2008.

MENEZES, R. S. C.; SALCEDO. I. H. Mineralization of N after incorporation of organic fertilizers in a Neossolo Regolitico cultivated with maize. **Revista Brasileira Engenharia Agricola e Ambiental**, Campina Grande, v.11, p.361-367, 2007.

MELO, L. C. A.; SILVA, C. A.; DIAS, B. O. Caracterizaçao da matriz orgânica de residuos de origens diversificadas. **Revista Brasileira de Ciências do Solo**, Viçosa, v. 32, n. 01, p. 101-110, 2008.

MOREIRA, F. M. S.; SIQUEIRA, J. O. **Microbiologia e bioquimica do solo**. Lavras: UFLA, 2006, 729p.

MUELLER, S. et al. Tomato productivity under organic fertilization and mineral fertilizer supplementation. **Horticultura Brasileira**, Brasilia, v. 31, n.1, p. 86-92, 2013.

PIRES, J. F.; JUNQUEIRA, A. M. R. Impact of organic fertilization on the productivity and quality of vegetables. **Horticultura Brasileira**, Brasilia, v. 19, n. 2, p. 195, 2001.

RAIJ, B. V. **Fertilidade do solo e aduçâo.** Piracicaba: Agronômica Ceres, Potatos. 1991, 343p.

RODRIGUES, G. O. et al. Amount of bovine manure in the agronomic performance of arugula (*Eruca sativa* L.), cultivated cultivar. **Revista Caatinga**, Mossoró, v. 21, n. 1, p. 162-168, 2008.

SALAZAR, F. J. et al. Nitrogen budgets for three cropping systems fertilized with cattle manure. **Bioresource Technology**, v. 96, p. 235-245, 2005.

SANTOS, H. C. **Sorption kinetics and phosphorus availability as a function of phosphorus contact time in the soil.** 2010, 56p. Thesis (Doctorate in Agronomy) - Agricultural Sciences Center. Federal University of Paraiba, Areia, 2010.

SILVA GALVÂO, S. R. SALCEDO, I. H.; OLIVEIRA, F. F. Accumulation of nutrients in sandy soils fertilized with cattle manure. **Pesquisa Agropecuària Brasileira**, Brasilia, v. 43, n. 1, p. 99-105, 2008.

SILVA, J. et al. Effect of cattle manure on yields of green ears and grains of maize. **Horticultura Brasileira**, Brasilia, v. 22, n. 2, p. 326-331, 2004.

SILVA, M. A. G. et al. Effect of potassium chloride on the salinity of a soil cultivated with pepper, *Capsicum annuum* L., in a protected environment. **Acta Scientiarum**, Maringà, v. 23, n. 5, p. 1085-1089, 2001.

SILVA, N. et al. Mineral and organic fertilization of hybrid pumpkin II. Nutritional status and production. **Pesquisa Agropecuària Tropical**, Goiânia, v. 29, p. 1928, 1999.

SIMÔES, M. L. et al. Characterization of organic fertilizers by electron paramagnetic resonance spectroscopy. **Revista Brasileira de Ciências do Solo**, Viçosa, v. 31, p. 1319-1327, 2007.

SOUZA, R. F. et al. Liming and organic fertilization: influence on phosphorus adsorption in soils. **Revista Brasileira de Ciências do Solo**, Viçosa, v. 30, n. 06, p. 975-983, 2006.

URQUIAGA, S.; MALAVOLTA, E. Urea: an organic fertilizer with potential for organic agriculture. **Cadernos de Ciências e Tecnologia**, Brasilia, v. 19, n. 2, p. 333-339, 2002.

CHAPTER I: PHYSIOLOGY AND GROWTH OF MELANCIE CULTIVATED UNDER DIFFERENT DOSES OF NPK

SOUSA - PB

2018

SUMMARY

OLIVEIRA FILHO, Francisco de Sales & PEREIRA, Francisco Hevilàsio Freire. **Physiology and growth of watermelon grown under different doses of NPK**, 2018. 71p.

The search for fertilizer alternatives that reduce or even eliminate the use of industrialized mineral fertilizers is a reality among small and medium-sized agricultural producers in the northeast of Brazil. The aim of this work was to evaluate the physiological and growth responses of watermelon to the application of doses of NPK using different proportions of mineral and organic fertilizers. The experiment was carried out at the Federal University of Campina Grande (UFCG), Campus Pombal - PB (6° 48" 16""S and 37° 49" 15""W), from June to September 2013. The design was in randomized blocks with four replications arranged in a 3x5 factorial scheme where, in factor (A) different concentrations of nutrients N, P and K were allocated (50, 100 and 150% of the NPK recommendation for watermelon) and in factor (B) five proportions of mineral and organic fertilizer (100/0, 75/25, 50/50, 25/75 and 0/100). The following were evaluated: gas exchange, dry mass of leaves, stem, fruit and total dry mass and leaf area. The concentration of 150% of the NPK recommendation for the watermelon crop was the most efficient in increasing the physiological characteristics and the accumulation of dry mass of the stem, leaf, fruit and total. The combined application of mineral and organic fertilizer provides a photosynthetic rate equivalent to the application of mineral fertilizer alone, with the 50/50 ratio being the most efficient for this variable. The use of manure alone does not contribute to the increase in dry mass in the vegetative part, when compared to its combined use with mineral fertilizers, with the 75/25 and 50/50 proportions being the most efficient in the accumulation of dry mass in the watermelon tree.

Keywords: *Citrullus lanatus*, gas exchange, manure, dry mass

1 INTRODUCTION

Fertilization is among the cultural practices that most contribute to the growth and development of crops. However, it is necessary to know about the dynamics of these materials in the soil and in the plant, in order to ensure that the nutrients applied can perform their specific functions in their most diverse sites of action.

In watermelon growing, mineral nutrition is an important factor that directly influences physiological and growth parameters. Nitrogen, potassium and phosphorus are the nutrients most commonly applied in fertilization and should be supplied according to the requirements of each cultivar, technological level, soil fertility, expected production, growth stage and climatic conditions (SOUZA, 2012). Efficient fertilizer management benefits the environment, due to lower levels of soil acidification, water eutrophication, groundwater pollution and salinization of areas, as well as benefiting various segments of society such as producers, technical agents and consumers (CARDOSO, 2011).

Among the representative parameters that can be used in the study of cultivated plants, growth analysis is an important and essential method for evaluating the behavioral differences of these plants, since they are influenced by agronomic practices and intrinsic factors associated with plant physiology (COSTA et al., 2006). Plant growth follows the dynamics of photosynthetic production and its analysis is vitally important for understanding the plant's morphological and physiological processes and their influence on yield (SOUZA, 2012). When measuring changes in plant growth, the accumulation of dry matter is the most significant parameter, since it results from the association of several other components (PEIXOTO et al., 2006).

The isolated application of industrialized mineral fertilizers is a reality among farmers, whether they are small, medium or large producers.

Several studies have been carried out in Brazil in order to observe the growth responses of olive trees to the application of various sources and doses of fertilizers (SILVA, 2010). Oliveira et al. (2009), when studying the influence of the application of increasing doses of N and K via mineral fertilizers and irrigation rates on the accumulation of dry mass in the melon tree, concluded that the increase in the doses of these macronutrients together with the increase in the irrigation rate increases the partitioning of dry mass of the vegetative part to the detriment of the dry phytomass of the fruit. Souza (2012) reported that different doses of N and P via mineral fertilizers influence the accumulation of dry mass in the vegetative part, fruit and leaf area of the watermelon tree, with extreme doses providing less total dry mass partitioning.

Research on the combined use of organic and mineral fertilizers and their effects on short-cycle crops such as watermelon is scarce. It can be seen that the few existing studies dealing with the use of organic and mineral fertilizers are based solely on the use of these materials in isolation or together, but there are no studies that consider the use of these fertilizers in different proportions and, in particular, that take into account the chemical and physical characteristics of organic fertilizers in order to quantify them more precisely, thus increasing the efficiency of their use, enabling farmers to make more practical and economical use of organo-mineral fertilizers.

The aim of this study was to evaluate the physiological and growth responses of watermelon as a function of fertilization with increasing doses of NPK applied via organic and mineral fertilizer in different proportions.

2. MATERIAL AND METHODS

The study was carried out from June to September 2013 in an area located at the Center for Agrifood Sciences and Technology of the Federal University of Campina Grande (CCTA/UFCG), situated in the municipality of Pombal, state of Paraiba, whose reference coordinates are 6° 48" 16" S latitude and 37° 49" 15" W longitude, at an altitude of 144 m. The climate of the region, according to the Koppen-Geiger classification, is of the Aw" type. The region's climate, according to the Koppen-Geiger classification, is of the Aw" type, i.e. hot and dry with summer and fall rains (semi-arid). The region's soils are classified as fluvial neosols.

Figure 1 Experimental area after transplanting watermelon seedlings at the Center for Agrifood Sciences and Technology of the Federal University of Campina Grande. CCTA/UFCG, Pombal - PB. 2013.

Source: Personal archive

In the municipality of Pombal - PB, the average temperature during the field experiment was 26.0°C, with a maximum of 28.2°C and a minimum of 25.6°C, the daily relative humidity was around 57.0%, the average rainfall was 52.1 mm and the average Eto was 8.1 mm per day (AESA, 2013).

Before the experiment was set up, chemical analyses of the soil were carried out at the Soil Fertility and Plant Nutrition Laboratory of the Universidade Federal do Semiàrido in Mossoró - RN, and the chemical characteristics are described in Table 1.

Table 1 Chemical characteristics of the soil **in** the 0-20 cm depth layer of **the** experimental area

Prof.	pH	P	K^-	Na^+	Ca^{2+}	Mg^{2-}	Al^{3-}	H^-+Al^{3+}	SB	CTC	V
Cm	H_2O	mg dm^{-3}				$cmol_c$ dm^{-3}					%
0-20	6,58	43,6	138,8	95,2	19,6	5,28	0,0	3,05	25,65	28,70	89

P, K. Na: Melichl Ejttrator ; Al_1 Ca,Mg: ExtiatarKCl IM; SB = Ca^+ ⅛Mg ^-K7+Na^+ ; H-Al: ExtiatarAcetato de Calcio 0,5 M_3 pH 7,0; CTC = SB+EΓ +Al^{++j} ; M OL: Ihgebtao UmidaWalkley-Black.

The treatments consisted of three percentages (50, 100 and 150%) of the doses of N, P2O5 and K2O recommended by Cavalcanti (2008) for watermelon, supplied via mineral and organic fertilizers, respectively, applied in different proportions (0:100, 75:25, 50:50 25:75 and 100:0). The 100% dose, based on the recommendation for watermelon growing and taking the soil's chemical analysis as a parameter, was 120 kg ha-1 for N, P2O5 and K2O. The experiment was set up using a randomized block design with the treatments distributed in a 3x5 factorial design, with four replications.

The mineral fertilizer sources used were monoammonium phosphate (MAP) (62% P2O5 and 12% N), urea (45% N) and potassium chloride KCl (60% K2O). The source of organic fertilizer used was dairy cattle manure, whose nitrogen, phosphorus and potassium contents are shown in Table 2. The amounts of manure and mineral fertilizers Urea, MAP and KCl calculated for the treatments are shown in Table 3.

Table 2 Chemical characteristics of the cattle herd

M.S.	N	P_2O_5	K_2O
dag dm^{-3}			
88	10,8	0,36	1,2

Nitrogen: Distillation-titration (Kjeldahl); Phosphorus: Spectrophotometry with molybdenum blue;

Potassium: Flame Photometry.

Table 3 Amounts of manure and mineral fertilizers in Kgha⁴₁ calculated for each treatment 50% OF RECOMMENDATION (Kg ha'¹)

TREATMENTS	Manure	Urea	MAP	KCl
Tl (IOflM : OOR)	0.00	108.33	100	104,16
T2(75Mz25OR)	**4500**	81.25	75	78.08
T3 (5 DM: 50OR)	9000	54,165	50	52,08
T4 (25 M : 75 OR)	13500	27,08	25	26,04
T5 (OM : 10 OOR)	18000	0=0	0,0	0,0

	100%	THE RECOMMENDATION		
T6 (IDOM : OOR)	0.00	216.66	200	208.33
T7 (75M = 25OR)	9000	162,5	150	156,25
TS (5 OM: 5 OOR)	18000	108,33	100	104,10
T9 (25M :75OR)	27000	54.165	50	52.08
TIO (OM : IOOOR)	36000	0,0	0,0	0,0
	150%	THE RECOMMENDATION		
Tll iIOOM:OOR)	0= 00	324,99	300	312,48
Tl 2 (75M:25OR)	20250	243.74	225	234.33
Tl 3 (50M:50OR)	27000	162,49	150	156,22
T14 (25M:75OR)	40500	81,25	75	71,108
Tl 5 (OMzIOOOR)	54000	0,0	0,0	0,0

OR: Organic Fertilizer: M: Mineral

The amount of organic fertilizer (dairy cattle manure), referring to 100% of the NPK recommendation, was defined according to the N-Total, P (P2O5) and K (K2O) contents present in the dry matter of the material. The amounts for the other percentages corresponding to the respective treatments were calculated from the 100% values. To calculate the amount of manure, we used the expression proposed by Furtini Neto et al. (2001) (Eq. 1) where, after calculating the amount of organic fertilizer as a function of the macronutrients N, P and K individually, we calculated the average, whose value was defined as 100% of the recommendation. The amounts of manure depending on the macronutrient content observed in this material were 5,263, 87,719 and 13,158 kg ha^{-1} respectively for the N, P and K content, and the average was 36,000 kg ha .$^{-1}$

Eq. 1: X = (A) /(B/100 x C/100 x D/100)

in which:

X = dose of organic fertilizer to be applied (kg/ha);

A = dose of N, P or K required by the crop for a given yield (kg/ha); B = dry matter content of the organic fertilizer (%);

C = content of N, P or K in the dry matter of the organic fertilizer (%);

D = conversion rate of N, P or K from organic to mineral form (30% for N and 50% for P and K).

Once the quantities of organic fertilizer for each treatment had been defined, it was distributed on the planting line and incorporated into the soil in one go, 15 days before transplanting, and then the trenches were built, the dimensions of which were: 0.20 m high, 0.45 m wide and 6 m long. After incorporating the manure, daily irrigation was started using drip lines with a spacing of 30 cm between emitters and a flow rate of 1.7 L per hour.

The mineral fertilizers were applied via fertigation using a venturi-type fertilizer injector, in installments throughout the crop cycle (Figure 2). The phosphate fertilizer (MAP) was applied in three installments, with the first application being made the day before transplanting and the others in the two weeks afterwards. Balancing was carried out to identify the amount of nitrogen applied via MAP and the missing amount was divided into eight applications throughout the crop cycle using urea as the fertilizer source. The potassium applied via KCl was distributed in ten applications, 10% in the foundation, 10% in the first two weeks (5% per week), 40% from the third to the sixth week (10% per week), 30% in the seventh and eighth week (15% per week) and 10% in the ninth and tenth week (5% per week).

Source: Archivopersonal

Figure 2 Incorporation of manure (A), raising of the beds (B)₁ irrigation system (C) and fertilizer injection by venturi (D). CCTAUFCGr Pombal-PB. 2013.

To apply the mineral fertilizers via fertigation, 16 mm registers were used at the beginning of each drip line per plot to control their application. The day before the mineral fertilizers were applied and after the area had been irrigated, the registers on the plots were closed, leaving only those corresponding to the first treatment to be applied the following day. After the application of a particular treatment, the records were closed and the records for the next treatment were opened. The interval between two applications was calculated according to the time needed to clean the irrigation system, taking into account the flow rate of the emitters.

Fertilization with the common micronutrients Ca, Mg and S was carried out for all treatments according to the recommendations for the crop (Table 4). Irrigation was managed based on an estimate of the daily reference evapotranspiration (ETo), which was obtained from climatic data from the semi-automatic weather station installed on site. The daily irrigation water was calculated to replace the crop evapotranspiration losses calculated for each stage of plant development. Soil preparation, other cultural treatments and plant health control were carried out in accordance with the needs and recommendations for watermelon growing (PUIATTI; SILVA, 2005).

Table 4 Amount of micro and macronutrients applied via fertigation for all treatments

Fertilizers	Formula	Quantity (Kg ha")[1]
Boric acid	$H_3 BO_3$	0,07176
Manganese sulphate	$MnSO_4$ 4HjO	0,03935
Zinc sulphate	$ZnSO_4$ **THjO**	0,00509
Copper sulphate	$CuSO_4$ **SH Q_2**	0.01736
Anion molybdate o	$(NH)_{4j}$ $MoTO_{24}$ 4H O_2	0.02S93
Magnesium sulphate	$MgSO_4$[1]	100
Calcium sulphate	CaSO $-_4$[1]	100

The watermelon plants were grown at a spacing of 2.0 x 0.60 m. The area of each experimental unit consisted of a 6 m row containing ten plants, eight of which were considered useful (Figure 3).

Source: Archivopersonal

Figure 3 Spacing between plants and between rows of watermelon plants 20 days after transplanting. CCTA UFCG, Pombal - PB, 2013.

The seedlings were grown in 128-cell Styrofoam trays using commercial substrate. Commercial watermelon seeds of the Olimpia hybrid were used, one per cell in order to avoid thinning and wasting seeds. The trays were kept in a greenhouse where they were irrigated daily by hand. The seedlings were transplanted when they had two well-formed leaves, 13 days after sowing. The watermelon seedlings were transplanted in the late afternoon, when transpiration is lowest, and then irrigated (Figure 4).

Source: Archivopersonal

Figure 4 Watermelon seedling production on flags (A and B) and transplantation of the seedlings to the experimental area (C and D). CCTA UFCG, Pombal - PB, 2013.

Weeds were controlled between the crop rows and between the plants by hand, using a hoe, twice during the cycle. Combing was carried out, which consists of moving the branches away from the area with the greatest wetness and the strips of land reserved for traffic. This operation was carried out three times before fruit set, as it not only facilitates weeding, spraying and harvesting, but also prevents the fruit from rotting due to contact with water or mechanical damage.

The physiological characteristics were evaluated at 45 days after transplanting (DAT), which corresponds to approximately 80% of vegetative growth. On this occasion, the photosynthetic rate (A), stomatal conductance (gs), transpiration (E) and intercellular CO2 concentration (Ci) were determined, measured using an infrared gas analyzer (IRGA) LCpro (Analytical Development, Kings Lynn, UK) with a constant light source of 1,200 μmol of photons m-2 s-1. The readings were taken on the fourth leaf of the main branch, counted from the apex, of one plant from the useful area per plot (Figure 5).

Source: Archivopersonal

Figure 5 Physiological analysis of watermelon leaves using an infrared gas analyzer (IRGA). CCTAUFCG Pombal- PB; 2013.

Growth analyses were carried out at the end of the crop cycle, which was 68 days after transplanting. First, the fruit was weighed and collected from six plants per plot. From these, the most representative fruit, whose weight was closest to the average of the plot and did not appear to be damaged or have any phytosanitary problems, was chosen for the drying sample. For the analysis of the vegetative part, two plants per experimental unit were collected and cut off close to the ground, at which point the total fresh mass of the leaves and stem were weighed separately. Samples of the fresh mass of the leaves and stems were then weighed, placed in paper bags and dried in a 70°C oven for 72 hours. From the dry mass of the samples, the dry mass per plant was calculated for stems, leaves and fruit. The total dry mass was obtained by adding up the dry mass of the stem, leaf and fruit. The leaf area, in cm2, was obtained using the disk method, which was determined by relating the dry mass of eight leaf disks of known area to the dry mass of the leaves. The leaf area was calculated using the equation: AF = (MSF x ADF)/MSD, where AF is the estimated leaf area, MSF is the total dry mass of the leaves, ADF is the known area of the disks removed from the leaves and MSD is the dry mass of the disks removed from the leaves.

The data was submitted to analysis of variance and the means were compared using the Tukey test at 5% probability. SAEG software, Version 9.1, was used.

3. RESULTS AND DISCUSSION

There was an individual effect for the factors NPK concentrations and the different proportions of mineral and organic fertilizer for photosynthesis (A), transpiration (E) and stomatal conductance (gs) (Figure 6). There was a significant interaction between the concentrations of NPK and the proportions of mineral and organic fertilizer only for intercellular CO_2 concentration (Ci) (Figure 7).

The highest photosynthetic rate (17.44 µmol m_2 s^{-1}) was observed in the treatment where the highest fertilizer concentration was applied (150% of the crop recommendation) (Figure 6A). With regard to the proportion of fertilizers, it was observed that when only mineral fertilizer was applied, the photosynthetic rate was 17.82 µmol m^2 s^{-1} , which was higher than the treatment where only organic fertilizer was applied, where the value observed was 15.2 µmol m_2 s^{-1} . The fact that the ion Cl^- is necessary for the water molecule breaking reactions in photosynthesis, by which oxygen is produced (CLARKE; EATON RYE, 2000), may explain the higher photosynthetic rate in the plants subjected to the higher concentration of nutrients applied via mineral fertilizers, since one of the sources used was KCl. The nutrient potassium is also characterized by being an activator of a large number of enzymes, being closely related to the processes of assimilation of carbon dioxide and N, favouring the formation of nitrogen compounds and in the synthesis, translocation and storage of sugars (MALAVOLTA; CROCOMO, 1982). In the treatment where half mineral fertilizer and half organic fertilizer were applied (50/50), the photosynthetic rate did not differ significantly from the treatment composed only of mineral fertilizers (Figure 6B).

There was no interaction between the study factors on watermelon transpiration, but there was a significant effect of the factors in isolation. With regard to nutrient concentration, the highest transpiration was observed (4.8 mmol m^2 s^{-1}) due to the higher concentration (150%), which differed significantly from the 50 and 100% concentrations, whose transpiration values were 4.42 and 4.19 mmol m^2 s^{-1} , respectively (Figure 6C). Transpiration follows the same trend as photosynthesis, considering that CO assimilation$_2$ is linked to the loss of water from the plant to the environment (ANDRADE JUNIOR et al., 2011). With regard to the proportions of fertilizer, it was clear that transpiration was significantly influenced by the different treatments, where the proportions 100/0 and 50/50 provided the highest transpiration rates (4.54 and 4.69 mmol m^2 s^{-1}) and the proportion 75/25 the lowest (4.06 mmol m^2 s^{-1}) (Figure 6D).

Stomatal conductance was significantly influenced only by the nutrient concentration factor, where the highest conductance was observed as a result of the highest concentration (150% of the N, P and K recommendations) (figure 6E and F). The increase in stomatal conductance provided by the 150% concentration compared to the 100% concentration (recommended for watermelon crops) was 31.81%.

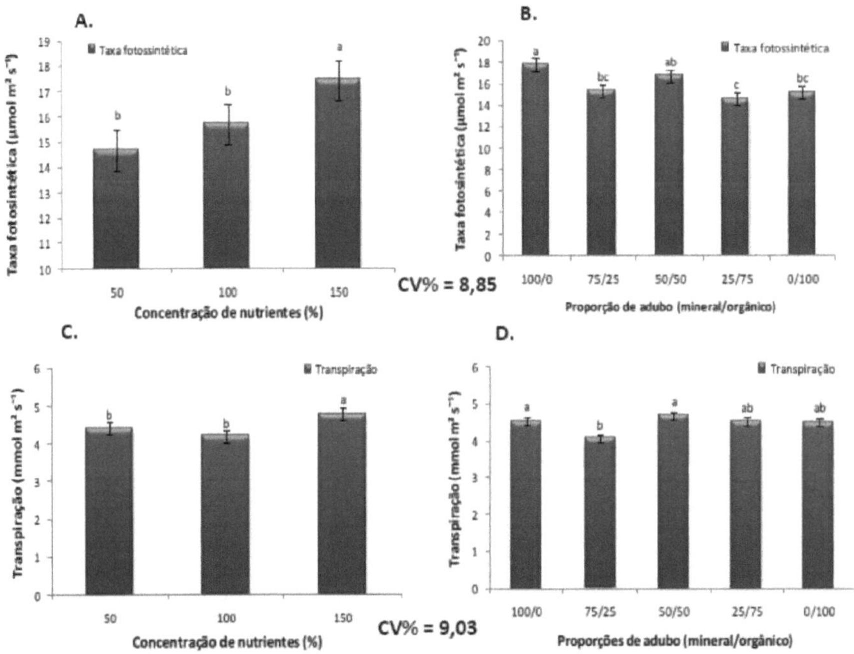

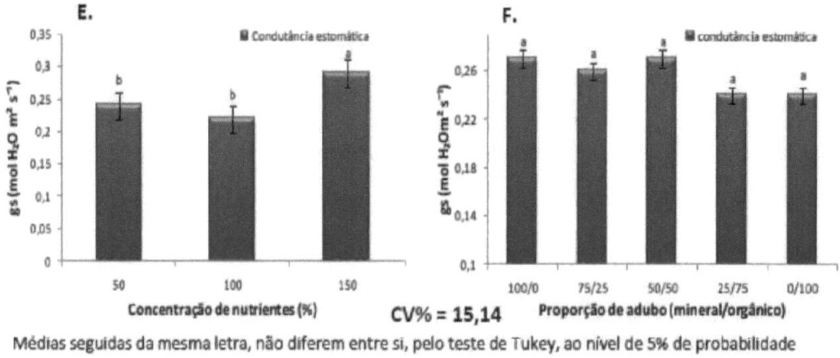

Figure 6 Photosynthetic rate (A and B)ı Transpiration (C and D) C□static conductance (E and F) in watermelon as a result of different concentrations of nutrients supplied via mineral and organic fertilizers. CCTATJFCGz Pombal - PBz 2013.

The higher stomatal conductance observed as a result of the higher concentration of nutrients may be directly related, in particular, to the greater availability of K^+ and Cl^-, since these elements are directly related to the opening of stomata in response to changes in the turgor of the guard cells (KERBAUY, 2008). It was observed that photosynthesis and transpiration were higher in the treatments that provided the greatest stomatal conductance. According to Koyro et al. (2013) photosynthetic rate is strongly correlated with stomatal conductance, transpiration and water use efficiency, which in turn are directly correlated with plant growth. These results reveal the possibility of reducing the use of mineral fertilizers in watermelon planting under the conditions in which the work was carried out, which has positive repercussions from an economic point of view, given the reduction in the costs of acquiring such inputs.

For the variable intercellular concentration of CO_2 (Ci), there was a significant interaction between the two study factors, where the lowest concentration of nutrients (50%) when applied in the proportions 100/0, 75/25, 50/50 and 25/75 (mineral/organic) favored the highest intercellular concentrations of CO_2 . Although statistically similar to the others, the 75/25 ratio within the 50% concentration provided the highest intercellular concentration of CO_2 (219.25 μmol mol^{-1}) and the 0/100 ratio the lowest (200 μmol mol^{-1}) (Figure 7). If Ci was high for a given treatment compared to the others, it means that the CO_2 reaching the mesophyll cells is not being fixed in the carboxylative phase of photosynthesis, possibly due to damage to its structure. No relationship was observed between gs and Ci, reinforcing the idea that the higher Ci at the lower concentration (50%) is related to low carbon assimilation in the carboxylative stage of photosynthesis.

The results observed are possibly related to the reduced availability of nutrients provided by these treatments, a condition that directly affects the photosynthetic process in both the photochemical and carboxylative stages. Nutrients such as N, a constituent of chlorophylls, P, a constituent of ATP, NADPH, RuBP and Rubisco molecules, and K, as an enzyme activator and osmoregulator, are indispensable in the photosynthetic process (TAIZ; ZEIGER, 2009).

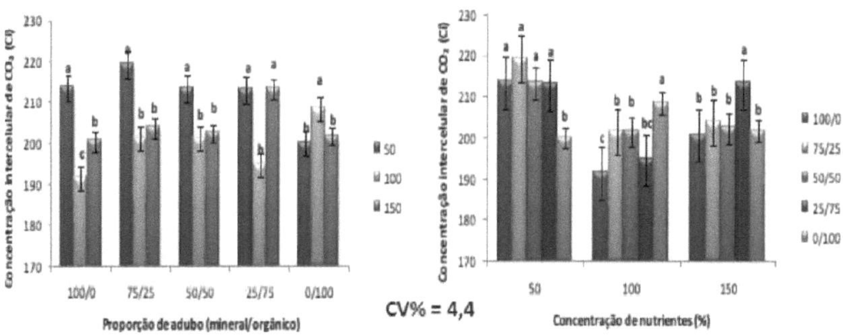

Averages followed by the same letter do not differ by Tukey's test at 5% probability **Figure** 7 Intercellular CO2 concentration in watermelon as a function of different nutrient concentrations applied via mineral and organic fertilizers. **CCTA/UFCG,** Pombal-PB. 2013.

There was an interaction between the factors nutrient concentration and fertilizer proportions for the variables leaf dry mass (LSM) and stem dry mass (SCM) (figure 8). There was an individual effect of the nutrient concentration factor for fruit dry mass (FDSM) (figure 8) and total dry mass (TDSM) (figure 9). The highest SDF accumulation was observed as a result of the highest nutrient concentration (150%) where the proportions 100/0, 75/25 and 50/50 were the ones that provided the highest mass accumulation in the leaves (110.88, 129.60 and 127.67 g plant^{-1} respectively) (Figure 8A and B). MSC followed the same trend as MSF, where the highest dry mass accumulations in the stem were 71.36, 70.47, 68.26 g plant^{-1} respectively for the 100/0, 75/25 and 50/50 ratios within the 150% concentration (Figure 8C and D).

It can be seen that the accumulation of SBM was higher than that of SCM. Work carried out with other cucurbits has found a certain similarity in the accumulation of mass in the different genera. Santos (2012) observed at the end of the melon cycle that, of the total dry mass accumulated in the vegetative part, the leaves accounted for 76.14% and the branches for 23.86%. The decrease in mineral fertilizer and the increase in organic fertilizer did not affect the accumulation of dry mass in the leaf and stem up to the 50/50 ratio between the two sources. For the 25/75 and 0/100 proportions, the low dry matter observed can be attributed to the slow availability of nutrients from the organic fertilizer, which according to Rodrigues et al. (2008), depends a lot on the degree of decomposition of the material used, and can have an immediate or residual effect. According to Sampaio et al. (2007), manure seems to cause immobilization of soil nutrients in the first month after its incorporation and the release after this period occurs progressively, reaching the highest quantities between three and six months after incorporation.

MSFR tended to be higher as nutrient concentrations increased, with values of 142.5, 191.5 and 242.0 g per plant, respectively, for the 50, 100 and 150% concentrations (N, P and K recommended for the crop) (Figure 8E). The different proportions of fertilizer did not significantly influence this variable (Figure 8F). The results show that the accumulation of dry mass in the fruit was greater than in the vegetative part. Similar results were observed by Grangeiro and Cecilio Filho (2004) in the watermelon hybrid Tide where, of the total dry mass accumulated during the cycle, the vegetative part was responsible for 29.6% and the fruit for 70.4%. Grangeiro and Cecilio Filho (2005) with the Shadow watermelon hybrid observed that at the end of the cycle the average contribution of the vegetative part to the dry mass accumulation was 34.4% and the fruit 65.6%. Aguiar Neto (2013) observed at the end of the cycle of the Quetzali cultivar in Petrolina and Mossoró, respectively, that the fruit accumulated 54 and 67% of the plant's total dry mass. Silva et al. (2012) observed that of the total dry mass accumulated by the watermelon tree, the fruit of the Olimpia cultivar accounted for 66% of the total dry mass.

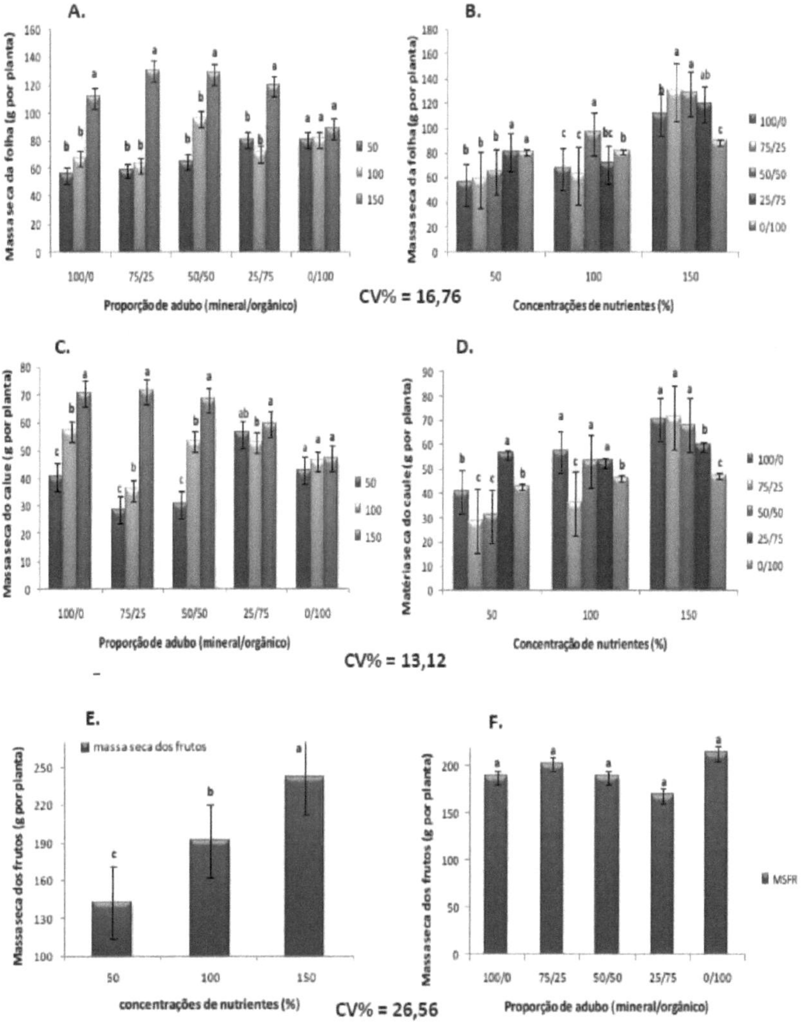

Figuin S Dry matter of the leaves (A and B), **stem** (C and D) and finto 5 (E and **F) of the** Hielancieira tree **as a** result of the different concentrations of nutrients applied via mineral and organic fertilization. CCTAdJFCG. Pombal - PB. 2013.

MST was calculated from the sum of the dry mass of the vegetative part (MSF, MSC and MSFR). The results confirm the above, where the highest concentration of nutrients (150%) provided the greatest accumulation of total dry mass (420.45 g per plant) (Figure 9A). It was observed that the use of organic fertilizer alone was not enough to achieve the highest accumulations of dry matter in the watermelon tree, but these were obtained with the application of chemical fertilizer alone or when combined with organic fertilizer as long as not less than 50% of mineral fertilizer was used. In view of the results, we can see that organic fertilizer was able to partially meet the watermelon tree's nutritional requirements, since when applied together with mineral fertilizer, the total dry mass accumulation was similar to treatments composed only of mineral fertilizers. The combination of mineral fertilizer with organic fertilizer may have improved the efficiency of nutrient absorption by the watermelon plants. Long-term studies comparing organic fertilization practices with conventional fertilization practices have shown greater dry mass accumulation in crops fertilized with organic materials (HERENCIA et al., 2011).

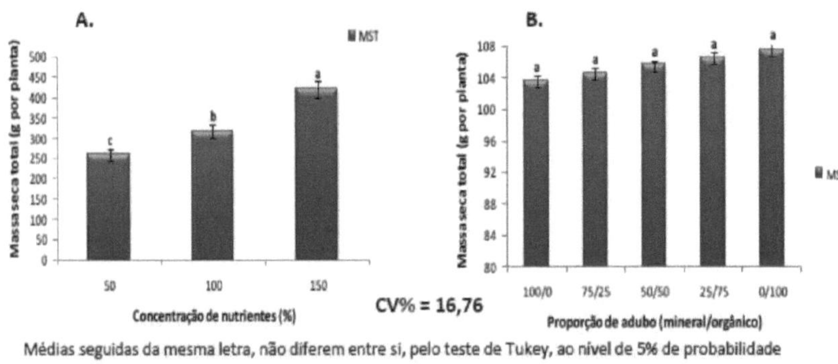

Fig. 9 Total dry matter (A and B) of watermelon as a function of different **nutrient** concentrations applied via mineral and organic fertilizers. CCTATJFCG$_3$ Pombal-PB$_3$ 2013.

For leaf area (LA), there was a significant interaction between nutrient concentration and fertilizer proportions (Figure 10). Within the proportions, the 150% concentration was the one that provided the greatest leaf area in watermelon plants. The 75/25 ratio (mineral/organic) provided the largest leaf area (29,613.03 cm^2) and the 0/100 the smallest (20,183.12 cm).^{2}

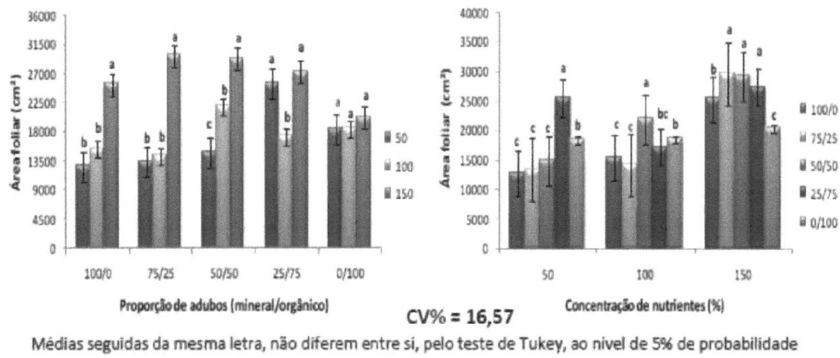

Figuтa 10 Area, foliar of melatici eira in riinçào of different concentrations of nutrients applied **via** ominerai and organic manure. CCTATJFCG$_{z}$ Pombal-PB$_{1}$ 2013.

In view of the results, it can be inferred that, under the conditions in which the experiment was carried out, the use of fertilizer recommendations used regionally may not be allowing the watermelon crop to express its maximum genetic potential, highlighting the importance of regional studies in order to increase the efficiency of fertilization practices. The methodology proposed in this work to quantify the manure to be used in different proportions with mineral fertilizers provided good results, showing that there is a possibility of adopting this practice at a local level.

4. CONCLUSIONS

The concentration of 150% of the NPK recommendation for the watermelon crop was the most efficient in increasing the physiological characteristics and the accumulation of dry mass in the stem, leaf, fruit and total;

The concentration of 100% of the NPK recommendation does not adequately meet the nutritional needs of the "Olimpia" watermelon crop in the Paraiba Sertao;

The combined application of mineral and organic fertilizer provides a photosynthetic rate equivalent to the application of mineral fertilizer alone, with the 50/50 ratio being the most efficient for this variable;

The use of manure alone does not contribute to the increase in dry mass in the vegetative part, when compared to its combined use with mineral fertilizers, with the 75/25 and 50/50 proportions being the most efficient in the accumulation of dry mass in the watermelon tree;

The optimum performance for the use of mineral and organic fertilizer sources in different proportions demonstrates the possibility of establishing cheaper fertilizer alternatives, especially for places where organic fertilizers are available at low cost.

BIBLIOGRAPHICAL REFERENCES

AESA (Executive Association of Water Management of the State of Paraiba) Agrometeorological Station of Pombal - PB, Available at <http://pcd.aesa.pb.gov.br/> Accessed 29/08/2013.

AGUIAR NETO, P. **Growth and nutrient accumulation in onions, melons and watermelons in the states of Rio Grande do Norte and Pernambuco.** 2013. 205p. Thesis (Doctorate in Agronomy: Plant Science) - Federal Rural University of the Semi-Arid (UFERSA), Mossoró-RN, 2013.

ANDRADE JUNIOR, et al. Effect of potassium nitrate on reducing salt stress in melon trees. **Revista caatinga**, Mossoró, v. 24, n.3, p. 110-119, 2011.

CARDOSO, S. M. **Sources and doses of nitrogen in the nutrition, production and quality of bean plants.** 2011. 65 p. Dissertaçao (Mestrado) - Universidade Estadual Paulista, Faculdade de Ciências Agronômicas - UNESP - FCA, Botucatu, 2011.

CAVALCANTE, F. J. A. **Recomendação de aduçâo para o estado de Pernambuco: 2ª aproximação**, Recife, IPA, 2008, 198p.

CLARKE, S. M.; EATON RYE, J. J. Amino acid deletions in loop C of the chlorophyll a-binding protein CP47 alter the chloride requirement and/or prevent the assembly of photosystem II. **Plant Molecular Biology**, v. 44, p. 591- 601, 2000.

COSTA, C. C. et al. Growth and assimilate partitioning in Cantaloupe melon as a function of phosphorus concentrations in nutrient solution. **Cientifica**, Jaboticabal, v. 34, n. 1, p. 123-130, 2006.

EMBRAPA. **Manual of soil analysis methods**. Rio de Janeiro: EMBRAPA Solos, 1997, 212p.

FILGUEIRA F. A. R. **Novo manual de olericultura**: Agrotecnologia moderna na produção e comercializaçao de hortaliças. 2ª ed. Revised and expanded. Viçosa: Federal University of Viçosa, 2003, 412p.

FURTINI NETO, A. E. et al. **Fertilidade do Solo**, Lavras, 2001. 252 p.

GONSALVES, M. V. I. **Doses of N and K applied via fertigation and plant spacing influencing development, productivity and water relations in seeded and seedless watermelon.** 2009, 98p. Dissertation (master's degree in agronomy - soil science) - Universidade Estadual Paulista "Julio de Mesquita Filho", Jaboticabal-SP, 2009.

GRANGEIRO, L. C.; CECILIO FILHO, A. B. Macronutrient accumulation and export by the watermelon hybrid Tide. **Horticultura Brasileira**, Brasilia, v.22, n.1, p.93-97, 2004.

GRANGEIRO, L. C.; CECILIO FILHO, A. B. Accumulation and export of nutrients by seedless watermelon, hybrid Shadow. **Cientifica**, Jaboticabal, v. 33, n. 1, p. 69-74, 2005.

HERENCIA, J. F. et al. Comparison of nutritional quality of the crops grown in an organic and conventional fertilized soil. **Scientia Horticulturae**, v. 129, p. 882888, 2011.

KERBAUY, G. B. **Fisiologia vegetal.** 2° ed. Rio de Janeiro, Ed. Guanabara Koogan, 2008, 431p.

KOYORO, H. W. et al. Photosynthetic and growth responses of a perennial halophytic grass *Panicum turgidum* to increasing NaCl concentrations. **Environmental and Experimental Botany**, v. 91, p. 22-29, 2013.

MALAVOLTA, E.; CROCOMO. O. J. Potassium and the plant. In: YAMADA, T.; IGUE, K.; MUZILLI, O.; USHERWOOD, N. R. **Potassium in Brazilian agriculture**. Piracicaba: IPF/IIP, 1982, p.95-162.

OLIVEIRA, F. S. et al. Accumulation and partitioning of dry matter, nitrogen and potassium by fertirrigated melon. **Bioscience Journal**, Uberlândia, v. 25, n. 3, p. 24-31, 2009.

PEIXOTO, C. P. et al. Growth analysis of different citrus genotypes grown under water deficit. **Revista Brasileira de Fruticultura**, Jaboticabal, v. 28, n. 3, p. 439-443, 2006.

PUIATTI, M.; SILVA, D. J. H. **Watermelon culture.** In: FONTES, P. C. R. (Ed.) **Olericultura: teoria e pràtica.** Viçosa (MG), 2005, p. 385-406.

RODRIGUES, G. O. et al. Amount of bovine manure in the agronomic performance of cultivated arugula (*Eruca sativa* L.). **Revista Caatinga**, Mossoró, v.21, n.1, p.162-168, 2008.

SAEG. **Statistical Analysis System**. Version 9.1. Viçosa: Arthur Bernardes Foundation, 2007.

SANTOS, F. G. B. **Analysis of growth, production and quality of cantaloupe melon in temporary protected cultivation with agrotextile in Mossoró-RN.** 2012. 137 f. Thesis (Doctorate in Plant Science) UFERSA, Mossoró, 2012.

SAMPAIO, E V. S. B.; OLIVEIRA, N. M. B.; NASCIMENTO, P. R. F. Efficiency of organic fertilization with cattle manure and *Egeria densa.* **Revista Brasileira de Ciência do Solo**, Viçosa, v. 31, p. 995-1002, 2007.

SILVA, M. V. T. et al. Fertilized watermelon growth under optimal nitrogen and phosphate fertilization conditions. **ACSA**, Patos, v. 9, n. 4, p. 61-66, 2012.

SILVA, J. S. **Evapotranspiration and watermelon production under different levels of nitrogen and irrigation water salinity**. 2010. 92 f. Dissertation (Master's Degree in Irrigation and Drainage) - Universidade Federal Rural do Semi-Àrido (UFERSA), Mossoró, 2010.

SOUZA, M. S. **Nitrogen and phosphorus applied via fertigation to watermelon hybrids Olimpia and Leopard.** 2012, 282p. Thesis (Doctorate and

Agronomy: Plant Science) - Federal Rural University of the Semi-Arid Region (UFERSA), Mossoró, 2012.

TAIZ, L.; ZEIGER, E. **Plant physiology**. 4ª . ed. Porto Alegre: Artmed, 2009. 819 p.

CHAPTER II: PRODUCTION, QUALITY AND NUTRIENT ACCUMULATION IN APPLE TREES GROWN UNDER DIFFERENT DOSES OF NPK

SOuSA - PB

2018

SUMMARY

OLIVEIRA FILHO, Francisco de Sales & PEREIRA, Francisco Hevilàsio Freire. **Production, quality and nutrient accumulation in watermelon grown under different doses of NPK**. 2018. 71p.

The search for fertilization alternatives that reduce or even eliminate the use of industrialized mineral fertilizers is a reality among small and medium-sized agricultural producers in the Northeast. The aim of this study was to evaluate the yield, fruit quality and nutrient accumulation of watermelon grown under different doses of NPK supplied by different proportions of mineral and organic fertilizer. The experiment was carried out in an area located at the Federal University of Campina Grande (UFCG), Campus Pombal - PB (6° 48" 16""S and 37° 49" 15""W), between May and July 2013. The design was in randomized blocks with four replications arranged in a 3x5 factorial scheme where, in factor (A) different concentrations of nutrients N, P and K were allocated (50, 100 and 150% of the NPK recommendation for watermelon) and in factor (B) five proportions of mineral and organic fertilizer (100/0, 75/25, 50/50, 25/75 and 0/100). The variables evaluated were: accumulation of nutrients in the leaves, stem, fruit and total, yield per plant, titratable acidity, soluble solids and the ratio between total soluble solids and titratable acidity. The concentration of 150% of the recommended dose was the most effective in the accumulation of NPK in the leaves, stem and fruit of the watermelon tree; the proportions of organic and mineral fertilizer 50/50 and 75/25 were more effective in the accumulation of N and P and 0/100 in that of K in the leaves of the watermelon tree; the accumulation of NPK in the fruit and vegetative part showed the following order of magnitude: K>N>P; the fruits were the preferential drains in NPK in relation to the vegetative part; the concentration of 150% of nutrients was more effective in increasing productivity in the proportions of mineral and organic fertilizers of 0/100, 75/25 and 25/75; The 100% and 150% concentrations were the most effective in the accumulation of total soluble solids when combined with the 75/25 and 50/50 proportions of mineral and organic fertilizer; the application of manure alone contributed to the greater accumulation of soluble solids in the fruit, especially when the NPK was applied in low concentrations to the soil.

Keywords: *Citrullus lanatus*, fertilizers, manure, mineral nutrition, yield

1 INTRODUCTION

Brazil occupies the fourth position in the ranking of the twenty largest watermelon producers in the world, but has the lowest yield in the group of countries that lead production (FAOSTAT, 2011). Watermelon production in Brazil in 2012 was 2,079,547 tons in 94,612 hectares of harvested area, which corresponds to an average yield of 21.97 tons ha^{-1} . The northeast region is responsible for 30.89% of the national production of the fruit and the states that stand out as the biggest producers are Bahia, Rio Grande do Norte and Cearà, respectively. Paraiba is among the northeastern states with the smallest area harvested for the fruit, 206 ha, and the lowest production, 4,002 t, ahead only of the state of Alagoas, which has a harvested area of 82 ha and an average production of 2,200 t (IBGE, 2012).

The majority of watermelon production in Brazil has a predominantly family production profile due to its rusticity, lower capital investment compared to other vegetables, and quick return of around 85 days (CARVALHO, 2005; FILGUEIRA, 2008).

The practice of fertilization contributes a significant amount to increasing the productivity and profitability of the crop, and proper fertilizer management is necessary in order to optimize production costs (NOBRE, 2007). The watermelon crop, like other olive crops, has mineral nutrition as one of the factors that directly contributes to productivity and fruit quality (BARROS et al., 2012). Increases in production, provided by the use of fertilizers, must be accompanied by an increase or maintenance of the quality of the fruit produced (GRANGEIRO; CECÌCILIO FILHO, 2004; MEDEIROS, 2008).

Most farmers make use of industrialized mineral fertilizers to supply nutrients to crops. However, in addition to increasing production costs, this practice has had undesirable effects on the environment, such as the pollution of reservoirs and watercourses and the salinization of soils in semi-arid regions. However, the rational use of inputs and natural resources, aimed at increasing agricultural production, are the two pillars of environmentally and economically viable agriculture, guaranteeing the sustainability of agro-ecosystems (FAQUIN et al., 2007).

The rising costs of mineral fertilizers, coupled with high losses in the field, require management practices that result in high efficiency of nutrient use by crops. Proper fertilizer management is crucial for a conscious and efficient production activity aimed at high yields and cost reduction, while respecting product quality and the environment.

The use of manure is a widely adopted alternative for supplying nutrients, mainly nitrogen and phosphorus, in family farming areas in the semi-arid and harsh regions of northeastern Brazil (MENEZES; SALCEDO, 2007). The use of organic waste as an alternative raw material for the production of fertilizers is a strategic measure from an environmental point of view, and is convenient as long as it is also viable from an economic point of view (FERNADES et al., 2003). Among the organic materials that can be used in agriculture, the use of animal manure occupies a prominent place with a positive effect on water infiltration and retention and an increase in the cation exchange capacity of soils (HOFFMANN et al., 2001). According to Kihel (1985), the use of organic fertilizers has a positive influence on production, even equaling or even surpassing the effects of mineral fertilization.

There are indications in the literature about doses of cattle manure to be used in watermelon growing. However, these studies have not demonstrated the effects of organic fertilization on watermelon production to justify these recommendations (LEÂO et al., 2008). Doses of organic fertilizers that maximize productivity and reduce costs, complemented by soluble mineral sources, could be established for regional conditions, with a view to improving the soil's physical and biological properties and reducing fertilizer costs (RODRIGUES; CASALI, 1999).

The aim of this work was to evaluate the influence of fertilization with increasing doses of NPK applied via organic and mineral fertilizer in different proportions on the components of production, the quality of watermelon fruit and the accumulation of nutrients in the watermelon tree in the semi-arid region of Paraíba.

2. MATERIAL AND METHODS

The experiment was set up from June to September 2013 in an area located at the Center for Agrifood Sciences and Technology of the Federal University of Campina Grande (CCTA/UFCG), situated in the municipality of Pombal, state of Paraiba, whose reference coordinates are: 6° 48" 16"" latitude S and 37° 49" 15"" longitude W, at an altitude of 144m. The region's climate, according to the Koppen-Geiger classification, is of the Aw" type, i.e. hot and humid with summer-autumn rains (semi-arid). The region's soils are classified as fluvial neosols.

In the municipality of Pombal, the average temperature during the field experiment was 26.0°C, with a maximum of 28.2°C and a minimum of 25.6°C, the relative humidity was around 57.0%, the average rainfall was 52.1 mm and the average Eto was 8.1 mm per day (AESA, 2013).

Before the experiment was set up, chemical analyses of the soil were carried out at the Soil Fertility and Plant Nutrition Laboratory of the Universidade Federal do Semiàrido in Mossoró - RN, and the chemical characteristics are described in Table 1.

Table 1 Chemical characteristics of the soil in the 0-20 m depth layer of the experimental area

Prof.	Ph	P	K^+	Na^+	Ca^{2+}	Mg^{2+}	Al^{3+}	$H^+ + Al^{3+}$	SB	CTC	V
Cm	H_2O	mg dm^{-3}			---------------cmol$_c$	dm^{-3}---------------					%
0-20	6,58	43,6	138,8	95,2	19,6	5,28	0,0	3,05	26,65	28,70	89

P, K, Na: Mehlich 1 Extractor; Al; Ca, Mg: KCl IM Extractor; SB = Ca* +M *$^2{}_g{}^1$ ÷K--Na*: H-Al: Calcium Acetate Extractor 0; 5 M, pH 7.0; CTC = SB+H*+A1^3 : MO.: Walkley-Black Wet Digestion.

The treatments were made up of three percentages (50, 100 and 150%) of the doses of N, P2O5 and K2O recommended by Cavalcanti (2008) for watermelon, supplied via organic and mineral fertilizers, respectively, applied in different proportions (0:100, 75:25, 50:50 25:75 and 100:0). The 100% dose, based on the recommendations for watermelon growing and taking the soil's chemical analysis as a parameter, was 120 kg ha^{-1} for N, P O_{25} and K_2 O. The experiment was set up using a randomized block design with the treatments distributed in a 3x5 factorial scheme, with four replications.

The mineral fertilizer sources used were monoammonium phosphate (MAP) (62%

P2O5 and 12% N), urea (45% N) and potassium chloride (KCl) (60% K_2O). The source of organic fertilizer used was dairy cattle manure, whose nitrogen, phosphorus and potassium contents are shown in Table 2. The amounts of manure and mineral fertilizers Urea, MAP and KCl calculated for the treatments are shown in Table 3.

The amount of organic fertilizer (dairy cattle manure), referring to 100% of the NPK recommendation, was defined according to the N-Total, P (P_2O_5) and K (K_2O) contents present in the dry matter of the material. The amounts for the other percentages corresponding to the respective treatments were calculated from the 100% values. To calculate the amount of manure, we used the expression proposed by Furtini Neto et al. (2001) (Eq. 1) where, after calculating the amount of organic fertilizer as a function of the macronutrients N, P and K individually, we calculated the average, whose value was defined as 100% of the recommendation. The amounts of manure depending on the macronutrient levels observed in this material were 5,263, 87,719 and 13,158 Kg ha^{-1} respectively for the N, P and K levels, and the average was 36,000 Kg ha^{-1}.

Eq. 1: X = (A) /(B/100 x C/100 x D/100)

in which:

X = dose of organic fertilizer to be applied (kg/ha);

A = dose of N, P or K required by the crop for a given yield (kg/ha);

B = dry matter content of the organic fertilizer (%);

C = content of N, P or K in the dry matter of the organic fertilizer (%);

D = conversion rate of N, P or K from organic to mineral form (30% for N and 50% for P and K).

Table 2 Chemical characteristics of cattle manure leiteiio

M.S.	N	P_2O_5	K_2O
	dag dm^{-3}		
88	10,8	0,36	1,2

Nitrogen: Distillation - titration (KjeldaH); Phosphorus: Spectrophotometry with molybdenum blue; Potassium: Flame photometry.

Table 3 Amounts of manure and mineral fertilizers in Kgha⁴₁ calculated for each plot 50% OF RECOMEhTDAQA (Kg ha)⁴

TRAITS	Manure	TJreia	MAP	KCl
Tl (IOOM : OOR)	0.00	108.33	100	104 16
T2 (75M: 25OR.)	**4500**	81,25	75	78.08
T3 (5OM: 5OOR)	9000	54,165	50	52,08
T4(25M: 75 OR)	13500	27,08	25	26,04
T5 (OM : IOOOR)	18000	0,0	0,0	0,0
	100%	D THE REC O MENDATION		
T6 (IOOM : OOR)	0.00	216.66	200	208.33
T7 (75Mι25OR)	9000	162,5	150	156,25
T8 (5 OM: 5OOR)	18000	108,33	100	**10410**
T9(25Mι75OR)	27000	54.165	50	52,08
TlO (OM: IOOOR)	36000	0 O3	0,0	0,0
	150%	D A REC O MENDATION		
Tll (100M:OOR)	THE OO3	324,99	300	312,48
Tl 2 (75Mι25OR)	20250	243.74	225	234.33
Tl 3 (50Mι5OOR)	27000	162,49	150	156,22
T14 (25M:75OR)	40500	81,25	75	78,108
Tl 5 (OMriOOOR)	54000	0,0	0,0	0,0

OR: Organic Fertilizer: M: Mining

Once the quantities of organic fertilizer for each treatment had been determined, it was distributed on the planting line and incorporated into the soil in one go, 15 days before transplanting, and then the trenches were built, the dimensions of which were: 0.20 m high, 0.45 m wide and 6 m long. After incorporating the manure, daily irrigation began using drip strips with a spacing of 30 cm between emitters and a flow rate of 1.7 L per hour.

The mineral fertilizers were applied via fertigation using a venturi-type fertilizer injector, in installments throughout the crop cycle. The phosphate fertilizer (MAP) was applied in three installments, with the first application being made the day before transplanting and the others in the two weeks afterwards. Balancing was carried out to identify the amount of nitrogen applied via MAP and the missing amount was divided into eight applications throughout the crop cycle using the fertilizer Urea as the source. The potassium applied via KCl was distributed in ten applications, 10% in the foundation, 10% in the first two weeks (5% per week), 40% from the third to the sixth week (10% per week), 30% in the seventh and eighth week (15% per week) and 10% in the ninth and tenth week (5% per week).

To apply the mineral fertilizers via fertigation, 16 mm registers were used at the beginning of each drip line per plot to control their application. The day before the mineral fertilizers were applied and after the area was irrigated, the plot registers were closed, leaving only those corresponding to the first treatment to be applied the following day. After the application of a given treatment, the records were closed and the records for the next treatment were opened. The interval between two applications was calculated according to the time needed to clean the irrigation system, taking into account the flow rate of the emitters.

Fertilization with micronutrients, Ca, Mg and S was common for all treatments, as recommended for the crop (Table 4). Irrigation was managed based on an estimate of the daily reference evapotranspiration (ETo), which was obtained from climate data from the semi-automatic weather station installed on site. The daily irrigation water was calculated to replace the crop evapotranspiration losses calculated for each stage of plant development. Soil preparation, other cultural treatments and plant health control were carried out in accordance with the needs and recommendations for watermelon growing (PUIATTI; SILVA, 2005).

Table 4 Amount of micro and hynchronutrients applied via fertigation for all treatments

Fertilizers	Formula	Quantity (Kg ha')[1]
Boric acid	$HjBO_j$	0,07176
Manganese sulphate	$MnSO_4$ 4H O_3	0,03935
Zinc sulphate	$ZnSO_4$ TH O_3	0,00509
Copper sulphate	$CuSO_4$ SH Q_3	0.01736
Anion molybdate o	(NH $)_{4j}$ $MoTO_{34}$ 4H O_3	0,02893
Magnesium sulphate	$MgSO_4$[1]	IOO
Calcium sulphate	CaSO $-_4$[1]	IOO

The watermelon plants were grown at a spacing of 2.0 x 0.60 m, with the useful area of each experimental unit consisting of a 6 m row containing eight useful plants.

The seedlings were grown in 128-cell Styrofoam trays using commercial substrate. Commercial watermelon seeds of the variety "Olimpia", cutivar Crimson Sweet, were used, one per cell in order to avoid thinning and wasting seeds. The trays were kept in a greenhouse where they were irrigated daily by hand. The seedlings were transplanted when they had two well-formed leaves, 13 days after sowing. The watermelon seedlings were transplanted in the late afternoon, when transpiration is lowest, and then irrigated.

Weeds were controlled between the crop rows and between the plants manually, using a hoe, as often as necessary to keep the crop from competing for water, light and nutrients. Combing was carried out, which consists of moving the branches away from the area with the greatest wetness and the strips of land reserved for traffic. This operation is carried out three times before fruiting, because as well as making weeding, spraying and harvesting easier, it prevents the fruit from rotting due to contact with water or mechanical damage.

For the production evaluation, six plants per plot were considered to be useful and only those fruits harvested weighing more than 4 kg that did not show any damage or phytosanitary problems were considered to be commercial. For the chemical analysis of the fruit, samples were taken from the pulp of a slice, taken lengthways, from the apex to the back end, and homogenized in a multiprocessor to obtain the juice. From this, the following characteristics were determined, according to the methodology of the Adolfo Lutz Institute (2008): soluble solids (SS), titratable acidity (AT), SS/AT ratio (Figure 2). Soluble solids were determined by refractometry with the results expressed as a percentage; for titratable acidity, 2 mL of the juice was pipetted into 50 mL of distilled water with two drops of phenolphthalein and titrated with 0.1 M sodium hydroxide solution under agitation until a persistent pink color was obtained for 30 s, with the results expressed as a percentage of citric acid; the SS/AT ratio was obtained by dividing the soluble solids content by the titratable acidity.

Fonte: Arquivo pessoal

Figure 1. Weighing the fruit (A), extraction of watermelon juice (B), determination of titratable acidity (C) and total soluble solids (%) (D). CCTA/UFCG, Pombal - PB, 2013.

Nutrient accumulation was determined after drying the leaves, stems and fruit collected at 68 DAT, where they were crushed and analyzed for macronutrient levels of nitrogen, phosphorus and potassium according to the methodology used by Silva (2009). The plant material was analyzed at the Soil and Plant Nutrition Laboratory of the CCTA/UFCG - Pombal - PB.

The data was subjected to analysis of variance and the means were compared using the Tukey test at a 5% probability level. SAEG software, Version 9.1, was used.

3. RESULTS AND DISCUSSION

There was a significant interaction between the factors NPK concentration and proportions of mineral and organic fertilizer on the accumulation of nitrogen (N), phosphorus (P) and potassium (K) in the stems of watermelon plants (Figure 2). The unfolding of the interaction between the two factors showed that, for all the fertilizer proportions, the highest concentration of nutrients (150%) was the one that gave the highest N accumulation, with the 50/50 proportion giving the highest value (0.73 g $plant^{-1}$) followed by the proportions (100/0 and 25/75) whose values were 0.67 and 0.64 g $plant^{-1}$ respectively (Figure 2A and 2B). A similar result was observed for the accumulation of P and K in the stem, where the 150% concentration in the different proportions also favored the greatest accumulation of these nutrients. For the 100/0 and 50/50 proportions, the amounts of P observed in the stem were similar, with a value of 0.36 g $plant^{-1}$ (Figure 2C and 2D).

In the case of K, the proportions 100/0, 25/75 and 0/100 in the 150% concentration gave the highest values (2.18; 2.12 and 2.11 g $plant^{-1}$ respectively) (Figure 2E and 2F). Based on the results about the greater export of N and P by the stem when only mineral fertilizer was used, this result can be attributed both to the fact that these fertilizers are highly soluble and readily available to the plant and to the fact that they were applied in parcels and in different quantities throughout the crop cycle, which may have favored their absorption at the different growth stages. The 25/75 ratio may have benefited the mineralization of the manure due to the supply of nutrients to the soil microorganisms, thus attenuating the process of nutrient immobilization and making the nutrients N and P more available for absorption by the watermelon plants. The incorporation of organic matter into the soil has been shown to rapidly adsorb phosphorus applied as fertilizer, increasing the availability of P (GUPPY et al., 2005). With regard to K, it can be deduced from the data that the similar export between the proportion where only mineral fertilizer was used and that with only organic fertilizer is due to the fact that this nutrient is not part of organic compounds (TAIZ and ZEIGER, 2009) and is easily released into the soil solution. In general, the decreasing order of nutrient accumulation for all treatments was: K>N>P.

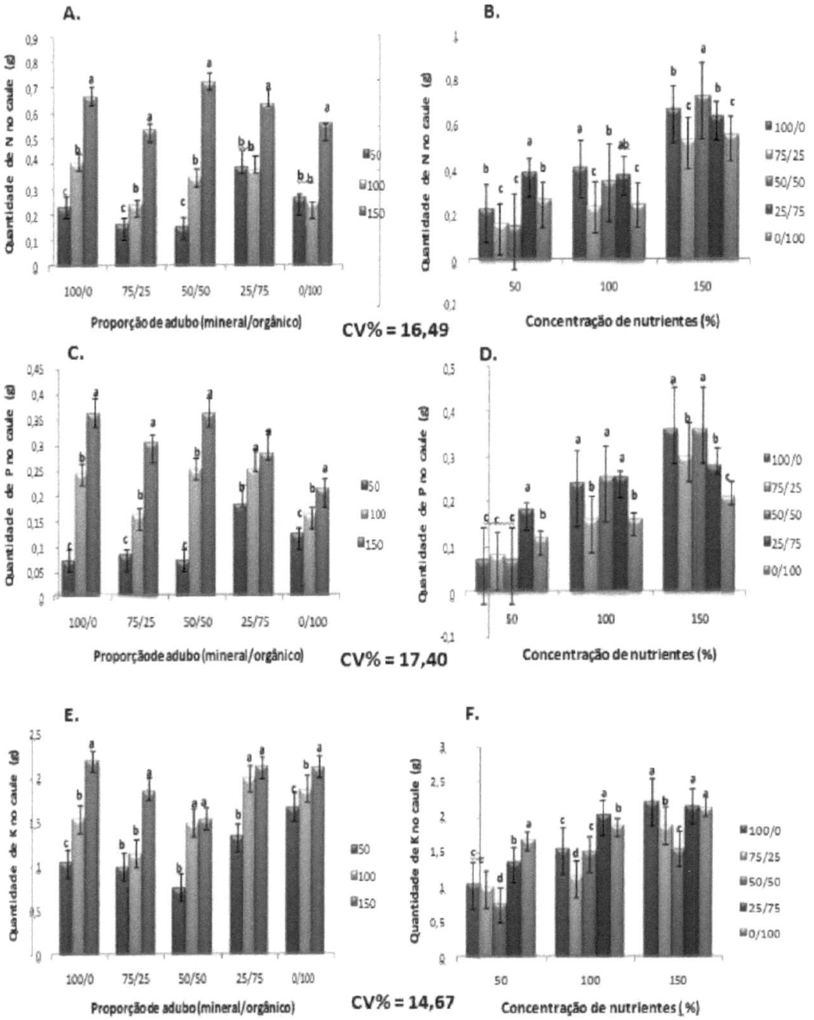

Médias seguidas da mesma letra, não diferem entre si, pelo teste de Tukey, ao nível de 5% de probabilidade

Figure 2: Amount of nitrogen (A and B), phosphorus (C and D) and potassium (E and F) in the stem of the watermelon tree as a function of different concentrations of nutrients applied via mineral and organic fertilization. CCTAXJFCG, Pombal - PB. 2013.

The accumulation of N, P and K in the leaves showed a significant interaction between the factors under study (Figure 3). The export of these nutrients by the leaves followed the same downward trend as the absorption by the stem: K>N>P. Grangeiro and

Cecilio Filho (2004), evaluating the accumulation and export of macronutrients by the watermelon hybrid Tide, observed similar results to this study.

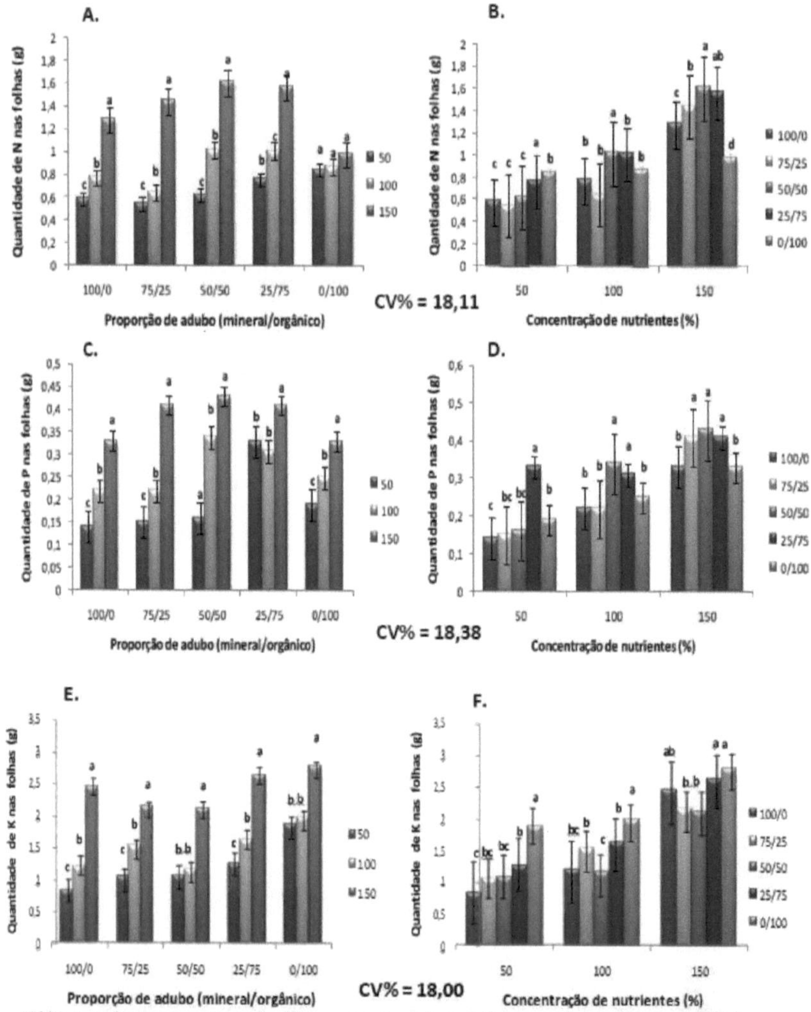

Médias seguidas da mesma letra, não diferem entre si, pelo teste de Tukey, ao nível de 5% de probabilidade

Figura 3. Amount of nitrogen (A and B), phosphorus (C and D) and potassium (E and F) in watermelon leaves as a function of different concentrations of nutrients applied via mineral and organic fertilization. CCTATJFCG Pombal - PB, 2013.

The 150% concentration favored the greatest accumulation of N and P in the leaves in all the proportions evaluated. However, the 75/25, 50/50 and 25/75 ratios were the ones that gave the highest values, which were 1.44, 1.61 and 1.57 g $plant^{-1}$ of N and 0.41, 0.43 and 0.41 g $plant^{-1}$ of P respectively (Figure 3A, B, C and D).

With regard to the accumulation of K in the leaves, a similar result was observed to the stem, where the highest accumulations were obtained at the highest concentration of nutrients (150%), applied in the proportions 100/0, 25/75 and 0/100, whose values were 2.44, 2.63 and 2.78 g $plant^{-1}$, respectively. It can also be seen that, even at the lowest concentrations, the proportions with the highest amounts of organic fertilizer provided the greatest accumulation of K in the leaves. In view of the results, it can be seen that the export of macronutrients by the leaves was higher than by the stem, especially the potassium nutrient, which can be explained by the fact that this nutrient plays very important roles in the physiological processes that take place in this organ, such as photosynthesis, enzyme activation, protein synthesis and carbohydrate transport, among others (TAIZ and ZEIGER, 2009).

With regard to the accumulation of macronutrients (N, P and K) in the watermelon fruit, there was no significant interaction between the factors studied. There was a significant individual effect only for the nutrient concentration factor (Figure 4). The accumulation of macronutrients was greater as a result of the increase in fertilizer concentrations. For N, there was an accumulation of 1.39, 1.89 and 2.82 g $plant^{-1}$, respectively, for the concentrations (50, 100 and 150%) (Figure 4A). With regard to P, a similar result was found to N, where the accumulation of P was 0.6, 0.78 and 1.09 g $plant^{-1}$ (Figure 4C). The same result was also seen for K accumulation, where the values were 2.82, 3.69 and 5.5 g $plant^{-1}$ (Figure 4E). The results show that the decreasing order of macronutrient absorption by the fruit follows the same trend as the vegetative part (K>N>P). In general, the fruit was responsible for the greatest export of macronutrients, accounting for 52.5% of the K, 54.6% of the N and 57.9% of the P exported. These values were arrived at by comparing the amounts of nutrients in the vegetative part with the fruit for the treatments that provided the greatest accumulations.

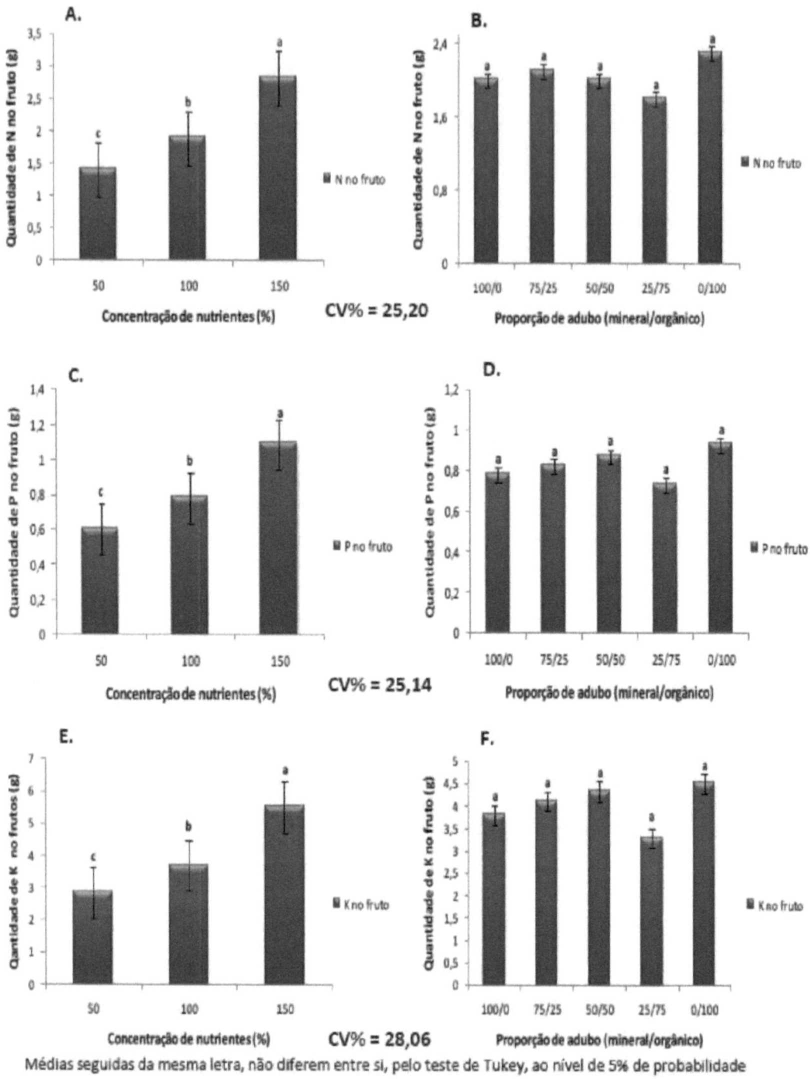

Figura 4. Amount of nitrogen (A and B), phosphorus (C and D) and potassium (E and F) in watermelon fruit as a function of different concentrations of nutrients applied via mineral and organic fertilization. CCTATJFCG; Pombal-PB, 2013.

For commercial watermelon production (Kg ha^{-1}), there was a significant interaction between the factors nutrient concentration and the ratio of mineral and organic fertilizers (figure 5). When the interaction was broken down, the highest yield was observed in the treatments subjected to the highest nutrient concentration of 150%

when applied in the proportions 100/0, 75/25 and 25/75 (mineral/organic). For the other proportions, there was no difference between the 100 and 150% concentrations (Figure 5A). For fertilization with manure alone (0/100), the yields in the 100 (57,500Kg ha^{-1}) and 150% (56,858Kg ha^{-1}) concentrations did not differ from each other. For the 100% concentration, the 50/50 and 0/100 proportions gave the highest yields. On the other hand, in the 150% concentration, the ratio that stood out was 75/25 (Figure 5B).

It is believed that the presence of organic fertilizer reduces losses through volatilization, leaching or immobilization, which may have contributed to the high watermelon yields obtained even with the higher proportions of manure. According to Pires and Junqueira (2001), as well as being an important source of nutrients, especially N, P, S and micronutrients, organic fertilizer is the only form of N storage that is not lost through volatilization and is also responsible for 80% of the total phosphorus found in the soil. There is a consensus among several authors on the efficiency of cattle manure associated or not with mineral fertilizers in increasing vegetable production. Fernandes et al. (2003) obtained higher melon yields as a result of applying organo-mineral fertilizers, with fruit production of 45.5 t ha^{-1} , higher than the 42.4 t ha^{-1} obtained with mineral fertilizers. Bertol et al. (2010) observed that the application of organic fertilizer compared to mineral fertilizer increased the concentration of total P, particulate P and reactive dissolved P in the soil surface.

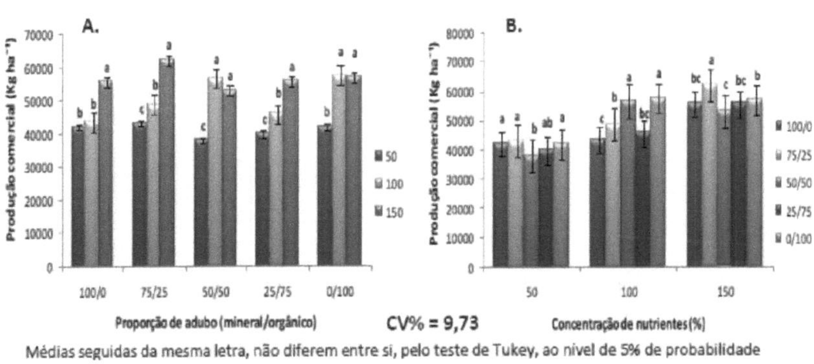

Figure 5 Commercial watermelon fruit production (A and B) as a function of different nutrient concentrations applied via mineral and organic fertilization. CCTA[Z] UFCG. Pombal- PB₅ 2013.

Leao et al. (2008), studying different levels of organic and chemical fertilization, observed lower watermelon yields (Crimson Sweet) when these were used alone. On the other hand, Araùjo et al. (1999), comparing the productive response of melon as a function of organic and mineral fertilization, observed a higher yield as a result of mineral fertilization and attributed the lower yield as a result of organic fertilization to the low level of K in the material used. Mueller et al. (2013) in their studies with organic fertilizer, with and without mineral fertilizer, in tomato cultivation observed higher commercial yields obtained with the application of mineral fertilizer alone or with the application of organic fertilizer supplemented with mineral fertilizer.

There was no significant effect for titratable acidity and the ratio of soluble solids to titratable acidity (SS/AT) as a function of the treatments evaluated. For the soluble solids content, there was a significant interaction between the nutrient concentrations and the different proportions of organic and mineral fertilizer (Figure 6). The highest SS contents were obtained in the 100% and 150% concentrations for all proportions, although they did not differ from the 50% concentration in the 100/0, 25/75 and 0/100 (mineral/organic) proportions (Figure 6A). The SS values observed were below the minimum level accepted by the consumer market according to the literature, which is 10% (BARROS et al., 2012). However, according to Leao et al. (2006), *there is* a varied spatial distribution of the SS content in watermelon pulp, being higher in the center of the fruit and gradually decreasing as it approaches the rind, which partly justifies the low levels observed due to the fact that it was determined in juice from a mixture of different parts of the pulp.

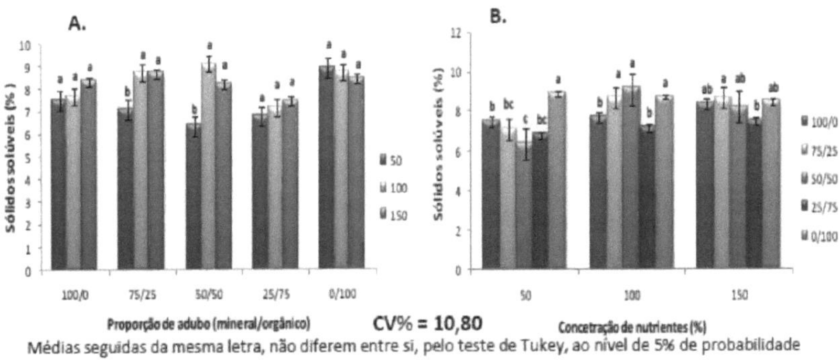

Figure 6 Soluble solids (A and B) in watermelon fruit as a function of different concentrations of nutrients applied via mineral and organic fertilization. CCTAXJFCG. Pombal - PB. 2013.

Fertilizing with manure alone led to a greater accumulation of SS in watermelon fruit at concentrations of 50 and 100% nutrients when compared to mineral fertilization alone, with increases of 18.6 and 13.3, respectively. In tomatoes, Polat et al. (2010) observed that organic fertilization resulted in higher fruit quality when compared to mineral fertilization, which should be maintained in order to facilitate the reuse and disposal of organic waste as well as to maintain and/or increase soil fertility.

4. CONCLUSIONS

The concentration of 150% of the recommended dose was the most effective in accumulating NPK in the leaves, stem and fruit of the watermelon tree;

The proportions of organic and mineral fertilizer 50/50 and 75/25 were more effective in the accumulation of N and P and 0/100 in K in watermelon leaves;

The accumulation of NPK in the fruit and vegetative part showed the following order of magnitude: K>N>P;

The fruit was the preferred drain on NPK compared to the vegetative part;

The 150% concentration of nutrients was more effective in increasing productivity in the mineral and organic fertilizer proportions of 0/100, 75/25 and 25/75;

The 100 and 150% concentrations were more effective in accumulating total soluble solids when combined with mineral and organic fertilizer proportions of 75/25 and 50/50.

The application of manure alone contributed to a greater accumulation of soluble solids in the fruit, especially when the NPK was applied in low concentrations to the soil.

REFERENCES

AESA (Associação Executiva de Gestao das Aguas do Estado da Paraiba) Agrometeorological Station of Pombal - PB, Available at <http://pcd.aesa.pb.gov.br/> Accessed 29/08/2013.

ARAÙJO, J. A. C.; GUERRA, A. G.; DURIGAN, J. F. Effects of organic and mineral fertilization on melon cultivars under greenhouse conditions. **Revista Brasileira de Engenharia e Ambiental**, Campina Grande, v. 3, n.1, p. 26-29, 1999.

BARROS, M. M. et al. Production and quality of watermelon subjected to nitrogen fertilization. **Revista Brasileira Engenharia Agricola e Ambiental**, Campina Grande, v.16, n. 10, p. 1078-1084, 2012.

BERTOL, O. J. et al. Phosphorus loss by surface runoff in no-till system under mineral and organic fertilization. **Scientia Agricola**, Piracicaba, v.67, n.1, p.71-77, 2010.

CARVALHO, R. N. **Growing Watermelons for Family Farming**. Brasilia: EMBRAPA, 2005, 112p.

CAVALCANTE, F. J. A. **Recomendação de aduçâo para o estado de Pernambuco: 2ª aproximação**, Recife, IPA, 2008, 198p.

FAOSTAT. **Agricultural Data**. 2012. Available at <http://faostat.fao.org/site/339/ default.aspx> Accessed on: 05/03/2014.

FAQUIN, V. et al. **Fertilizers and the Environment**. Lavras, UFLA/FAEPE, 2007, 86p.

FERNANDES, A. L. T.; RODRIGUES, G. P.; TESTEZLAF. Mineral and organomineral fertirrigation in relation to quality greenhause cultivated melon. **Scientia Agricola**, Piracicaba, v. 60, n.1, p. 149-154, 2003.

FILGUEIRA, F. A. R. **Novo manual de olericultura: agrotecnologia moderna na produçâo e comercializâo de hortaliças**. Viçosa: UFV, 2008, 402p.

FURTINI NETO, A. E. et al. **Fertilidade do Solo**. Lavras, 2001, 252 p.

GRANGEIRO, L. C.; CECILIO FILHO, A. B. Accumulation and export of nutrients in seedless watermelon. **Cientifica**, Jaboticabal, v.33, n.1, p.69-74, 2004.

GUPPY, C. N. et al. Competitive sorption reactions between phosphorus and organic matter in soil: a review. **Australian Journal of Soil Research**, Australia, v. 43, n. 2, p. 189-203. 2005.

HOFFMANN, I. A. et al. Farmers management strategies to maintain soil fert ility in a remote area in northwest Nigeria. **Agriculture, Ecosystems & Enviroment**, v. 86, n. 03, p. 263-275, 2001.

IAL - Adolfo Lutz Institute. **Physico-chemical methods for food analysis**. Sao Paulo: Instituto Adolfo Lutz, 1st ed. 2008, 1020p.

BRAZILIAN Institute of Geography AND STATISTICS. **Municipal agricultural production**. Temporary and permanent crops (2011). Rio de Janeiro, 2012. 97p.

KIEHL, E. J. **Organic fertilizers.** Piracicaba: Agronômica Ceres, 1985. 494p.

LEÂO, D. S. S.; PEIXOTO, J. R.; VIEIRA, J. V. Lycopene content and total soluble solids in eight watermelon cultivars. **Bioscience Journal**, Uberlândia, v. 22, n. 3, p. 7-15, 2006.

LEÂO, D. S. S. et al. Watermelon productivity under different levels of chemical and organic fertilization. **Bioscience Journal**, Uberlândia, v. 24, n. 4, p. 32-41, 2008.

MEDEIROS, D. C. **Production and quality of watermelon fertigated with N and K in Mossoró**, 2008. 70p. Thesis (Doctorate in Plant Science) Universidade Federal Rural do Semi- Àrido, Mossoró, 2008.

MENEZES, R. S. C.; SALCEDO. I. H. Mineralization of N after incorporation of organic fertilizers in a Neossolo Regolitico cultivated with maize. **Revista Brasileira de Engenharia Agricola e Ambiental**, Campina Grande, v.11, p.361367, 2007.

MUELLER, S. et al. Tomato productivity under organic fertilization and mineral fertilizer supplementation. **Horticultura Brasileira**, Vitória da Conquista, v. 31, n.1, p. 86-92, 2013.

NOBRE, J. G. A. **Responses of castor bean to irrigation and potassium application**

in red-yellow clay loam. 2007. 75p. Dissertation (Master's degree in Agronomy/Irrigation and Drainage) - Universidade Federal do Cearà, Fortaleza, 2007.

PIRES, J. F.; JUNQUEIRA, A. M. R. Impact of organic fertilization on the productivity and quality of vegetables. **Horticultura Brasileira**, Brasilia, v. 19, n. 2, p. 195-202. 2001.

POLAT, E.; DEMIR, H.; ERLER, F. Yield and quality criteria in organically and conventionally grown tomatoes in Turkey. **Scientia Agricola**, Piracicaba, v.67, n.4, p.424- 429, 2010.

PUIATTI, M.; SILVA, D. J. H. **Watermelon culture.** In: FONTES, P. C. R. (Ed.) **Olericulture: theory and practice.** Viçosa (MG), 2005, p. 385-406.

RODRIGUES, E. T.; CASALI, V. W. D. Yield and nutrient concentration in lettuce as a function of organic and mineral fertilization. **Horticultura Brasileira**, Brasilia, v. 17, n. 2, p. 125-128, 1999.

SILVA, F. C. **Manual de anâlises quimicas de solos, plantas e fertilizantes**. 2° ed. revisada e ampliada - Brasilia, DF: Embrapa Informaçao Tecnològica, 2009, 627 p.

TAIZ, L.; ZEIGER, E. **Plant physiology**. 4ª ed. Porto Alegre: Artmed, 2009. 819 p.

5 FINAL CONSIDERATIONS

This work provides important information about the possibilities for developing a more sustainable agriculture, less dependent on industrialized inputs and with less impact on the environment. Efficiency in the use of by-products generated in agricultural units is the focus of discussions between researchers in the search for agricultural production technologies that are less dependent on the dominant energy matrix, and on distant input markets, which will enable a greater increase in producer income and develop the economy at regional level.

The results of this study show that it is possible to establish fertilizer alternatives that are more economically viable and have less impact on the environment, especially in areas where organic materials are available at low cost.

The practice of fertilization is very complex, especially when working with organic fertilizers, given the heterogeneity of these materials and their multiple reactions with the edaphic environment, making it essential to carry out long-term regional studies to obtain more accurate results. Our study will therefore serve as a basis for further research to improve the technique of organo-mineral fertilization.

APPENDIX

Analysis of variance of the physiological, growth, nutritional, production and quality characteristics of melon fruit:

Booing[1]		A	E	gs	Ci
Source of Variation	GL	F			
Blocks	3	1.02ns	1.29ns	0.03ns	1 ,00ns
Proportion	4	2,88*	3,34**	1.98ns	0.99ns
Concentration	2	3,51**	2,32*	3,14*	*g* 32**
Interaction (P x C)	8	1.48ns	1.54ns	1.98ns	2,45*
Waste	42	-	-	-	-
Total	59	-	-	-	-
CV (%)		29,00	12,19	18,96	12,19

[1]Photosynthetic rate (A); transpiration (E); stomatal conductance (gs); intercellular CO concentration2 (Ci), ns, not significant; * sign, at 5º /o probability; ** sign, at 1% probability.

Variables[1]		MSF	MSC	MSFR	MST	AF
Source of Variation	GL	F				
Blocks	3	0.48ns	0.93ns	0.73ns	1.00ns	0.48ns
Proportion	4	6,27**	7,85**	1.31ns	0.22ns	6,27**
Concentration	2	48,63**	64,89**	19,03**	44,09**	48,63**
Interaction (P x C)	8	y yy**	9,49**	0.48ns	1.92ns	y yy**
Waste	42	-	-	-	-	-
Total	59	-	-	-	-	-
CV (%)		16,57	13,12	26,56	16,76	16.57

[1]Leaf dry mass (LSM); stem dry mass (SCM); fruit dry mass (FDM); total dry mass (TDM), leaf area (LA), ns, not significant; * sign, at 5% probabillty; ** sign, at 1% probability.

Variables[1]		ANC	APC	AKC
Source of Variation	GL	F		
Blocks	3	Lllns	1.45ns	1.39ns
Proportion	4	10,60**	9,75**	19,32**
Concentration	2	188,89**	148,46**	61,82**
Interaction (P x C)	8	4,29**	6,96**	3,32**
Waste	42	-	-	-
Total	59	-	-	-
CV (%)		16,49	17.40	11,67

[1]Accumulation of nitrogen in the stem (ANC); accumulation of phosphorus in the stem (AFC); accumulation of potassium in the stem (AKC). ns, not significant; * sign, at 5% probability; ** sign, at 1% probability

Variables[1]		.LNF	.LPF	AKF
Source of Variation	GL	F		
Blocks	3	0.60ns	0.03ns	0.30ns
Proportion	4	5,56**	10,35**	17,02**
Concentration	2	81,91**	68,12**	63,75**
Interaction (P x C)	8	5 44**	3,16**	4,72**
Waste	42	-	-	-
Total	59	-	-	-
CV (%)		18,11	18,38	18,00

[1]Nitrogen accumulation in the leaf (ANF); phosphorus accumulation in the leaf (AFF); potassium accumulation in the leaf (AKF). ns, not significant; * sign, at 5% probability; ** sign, at l° /o probability

Variables[1]		ANFR	APFR	AKFR
Source of Variation	GL	F		
Blocks	3	0.96ns	0.97ns	LOlns
Proportion	4	1.73ns	1.73ns	2.21ns
Concentration	2	39,62**	28,47**	29,54**
Interaction (P x C)	8	0.81ns	1.48ns	1.37ns
Waste	42	-	-	-
Total	59	-	-	-
CV (%)		25,20	25,14	28,06

[1]Nitrogen accumulation in the fruit (ANFR); phosphorus accumulation in the fruit (AFFR); potassium accumulation in the fruit (AKFR). ns, not significant; * sign, at 5% probability; ** sign, at 1% probability

Variables[1]		PROD.	SS	AT	SS/AT
Variation Factor	GL	F			
Blocks	3	2,71*	2.62ns	0.46ns	0.49ns
Proportion	4	2,76*	5.06**	1.04ns	2.40ns
Concentration	2	53,52**	7,67**	0.40ns	0.61ns
Interaction (P x C)	8	3,51**	2,41*	1.23ns	1.05ns
Waste	42	-	-	-	-
Total	59	-	-	-	-
CV (%)		9,73	10,80	19,52	29.00

[1]Yield (Yield); soluble solids (SS); titratable acidity (TA); ratio between soluble solids and titratable acidity (SS/AT), ns, not significant; * sign, at 5% probability; ** sign, at 1% probability

More Books!

yes

I want morebooks!

Buy your books fast and straightforward online - at one of world's fastest growing online book stores! Environmentally sound due to Print-on-Demand technologies.

Buy your books online at
www.morebooks.shop

Kaufen Sie Ihre Bücher schnell und unkompliziert online – auf einer der am schnellsten wachsenden Buchhandelsplattformen weltweit! Dank Print-On-Demand umwelt- und ressourcenschonend produzi ert.

Bücher schneller online kaufen
www.morebooks.shop

info@omniscriptum.com
www.omniscriptum.com

OMNIScriptum

MIX
Papier aus verantwortungsvollen Quellen
Paper from responsible sources
FSC® C105338

Printed by Books on Demand GmbH, Norderstedt / Germany